T0260893

CORAL WHISPERERS

CRITICAL ENVIRONMENTS: NATURE, SCIENCE, AND POLITICS

Edited by Julie Guthman, Jake Kosek, and Rebecca Lave

The Critical Environments series publishes books that explore the political forms of life and the ecologies that emerge from histories of capitalism, militarism, racism, colonialism, and more.

1. *Flame and Fortune in the American West: Urban Development, Environmental Change, and the Great Oakland Hills Fire*, by Gregory L. Simon

2. *Germ Wars: The Politics of Microbes and America's Landscape of Fear*, by Melanie Armstrong

3. *Coral Whisperers: Scientists on the Brink*, by Irus Braverman

CORAL WHISPERERS

SCIENTISTS ON THE BRINK

Irus Braverman

 UNIVERSITY OF CALIFORNIA PRESS

University of California Press, one of the most distinguished
university presses in the United States, enriches lives around
the world by advancing scholarship in the humanities, social
sciences, and natural sciences. Its activities are supported by
the UC Press Foundation and by philanthropic contributions
from individuals and institutions. For more information, visit
www.ucpress.edu.

University of California Press
Oakland, California

Library of Congress Cataloging-in-Publication Data

Names: Braverman, Irus, 1970– author.
Title: Coral whisperers : scientists on the brink /
    Irus Braverman.
Description: Oakland, California : University of California
    Press, [2018] | Includes bibliographical references and
    index. |
Identifiers: LCCN 2018008708 (print) | LCCN 2018013586
    (ebook) | ISBN 9780520970830 (epub and ePDF) |
    ISBN 9780520298842 (cloth : alk. paper) |
    ISBN 9780520298859 (pbk. : alk. paper)
Subjects: LCSH: Marine scientists—Interviews. | Coral reef
    management. | Coral bleaching—21st century.
Classification: LCC GC30.A1 (ebook) | LCC GC30.A1 B73 2018
    (print) | DDC 577.7/89—dc23
LC record available at https://lccn.loc.gov/2018008708
https://lccn.loc.gov/2018008708

Manufactured in Canada

26   25   24   23   22   21   20   19   18
10   9   8   7   6   5   4   3   2   1

For Ruth Gates

The publisher and the University of California Press Foundation gratefully acknowledge the generous support of the Ralph and Shirley Shapiro Endowment Fund in Environmental Studies.

CONTENTS

ACKNOWLEDGMENTS

I have many nonhumans and humans to thank for their support and encouragement in writing this book. I will start by thanking corals for bringing wonder into my days and for inspiring me to write about the scientists who study them. The first coral I recognized by name is *Acropora cervicornis*, a threatened coral I saw in the Caribbean nurseries when I first dove there in 2015. In the twenty-five years prior to that event, I didn't feel the need to name and classify underwater life—diving was a practice of being in the moment.

But each time I had the opportunity to dive with some of my interviewees, I discovered something new about marine ecology. To my surprise, rather than take away the wonder, this learning process enhanced it for me—and for this I have to thank each one of the coral scientists I interviewed. Every interview opened up a window to yet another aspect of what was otherwise an inaccessible world of ocean science. One hundred interviewees from different corners of the world are inevitably a varied bunch. While this variability was apparent on many fronts, one thing was true for all of the scientists, activists, and managers I interviewed: their deep care and passion for corals. The drive to save these simple yet complex organisms who are both so alien to us and yet also so familiar is nothing but admirable. Indeed, I often found myself envying this community, with its strong sense of commitment and belonging, and even considered taking a break from my position as professor of law and geography to pursue a Ph.D. in marine biology. On other days—especially around the 2016 U.S. election and its aftermath, when the gloom seemed unbearable—I was glad I had not chosen to be a marine biologist. How would I have dealt with life on the front lines of the slow-motion catastrophe for oceans coupled with the increasingly grim state of American politics? I wasn't sure I wanted to find out.

I would like to take a moment to thank the coral scientists who have let me into their world and who have shared with me their hopes and fears, their doubts and commitments, and their everyday research and debates. Specifically, I am grateful to Edwin Hernández-Delgado from the University of Puerto Rico, who welcomed me to his nursery in Culebra and was my first contact in this community. My gratitude extends to Howie Lasker and Mary Alice Coffroth of the University at Buffalo, who dedicated much time and energy to explaining coral biology and ecology to me and who have introduced me to many of their colleagues. Baruch (Buki) Rinkevich of the National Institute of Oceanography in Israel was always a fountain of ideas—it was a great pleasure to brainstorm with him for hours on end. Ruth Gates, director of the Hawai'i Institute of Marine Biology at the University of Hawai'i, has been a powerful source of inspiration and support. I would like to thank Murray Roberts of the University of Edinburgh for introducing me to the community of deep-sea coral scientists, and Chris Voolstra of the King Abdullah University of Science and Technology for his much-needed and very patient explanations of coral genomics.

I am also indebted to Iliana Baums of the Pennsylvania State University and the Center for Marine Science and Technology, Lorenzo Bramanti of the French National Centre for Scientific Research, Mark Eakin of NOAA's Coral Reef Watch, Nicole Fogarty of Nova Southeastern University's Oceanographic Center, Zac Forsman of the Hawai'i Institute of Marine Biology, Kristina Gjerde of the International Union for Conservation of Nature, Ben Halpern of the University of California Santa Barbara, Ove Hoegh-Guldberg of the University of Queensland and the Global Change Institute, Jeremy Jackson of the Scripps Institution of Oceanography and the Smithsonian Tropical Research Institute, Les Kaufman of Boston University, Ángela Martinez Quintana of the University at Buffalo, Margo McKnight of the Florida Aquarium, Stuart Newman of New York Medical College, and Peter Sale of the University of Windsor. These scientists generously read extensive parts of the manuscript and provided helpful comments, for which I am grateful.

I am also grateful to several of my nonscientist colleagues for reading and commenting on the manuscript: David Delaney, who carefully engaged with the early drafts of my work and encouraged me to stick with my unusual format; Jack Schlegel, who diligently commented on every version of the book, and always with much humor; and Guyora Binder, who provided prudent advice regarding the book's political and ethical dimensions. I would also like to thank Adam Rome and Carrie Bramen for their comments on the Introduction and to Errol Meidinger and Doug Kysar for their comments on Chapter 4. Parts of the manuscript greatly benefited from work-in-progress workshops at the Ecole Normale Supérieure in Paris and at the Rachel Carson Center for Environment and Society in Munich, both in June 2017, as well as at the Microbial Aesthetics Symposium, held at the University at Buffalo in November 2017. The Baldy Center for Law and Social Policy provided invaluable grants that enabled me to travel to Puerto Rico, Australia, and the Red Sea and helped fund my fieldwork in Hawai'i. I am also thankful for the support of the William J. Magavern Fellows Fund, for the University at Buffalo's OVPRED/HI Seed Money in the Arts and Humanities, and for the summer 2017 writing fellowship at the Rachel Carson Center for Environment and Society, which provided me both the physical and the mental space to complete the book. My research assistants John DiMaio, Eric Vanlieshout, and Miranda K. Workman helped me transcribe the interviews, conduct research, and edit the manuscript, and I thank each of them, as well as University at Buffalo librarian John Mondo, for their dedicated work. My editor from University of California Press, Kate Marshall, and her excellent staff—editorial assistant Bradley Depew, project editor Kate Hoffman, and copyeditor Richard Earles—marshalled the book through production with impressive skill. I am extremely grateful to the press's fantastic reviewers, Stefan Helmreich, Carrie Friese, and Peter Sale, as well as to the coeditor of the Critical Environments series, Julie Guthman. Colin Foord of Coral Morphologic and aquarist Julian Sprung provided invaluable assistance with many of the book's images.

Finally, I would also like to thank my older daughter, Ariel, for joining me on and helping me document so many of the fieldwork interviews and observations. Without her, I probably wouldn't have dared to fly on those tiny airplanes to small Caribbean islands or to dive again and again in the unpredictable Hawaiian waters. My little one, Tamar, has asked me the most thoughtful questions about coral biology, and helped me get through E. O. Wilson's work on ants and colonialism in animals. My partner, Gregor Harvey, has been holding it all together for all of us. I owe this book to these wonderful creatures whom I call my family, and also to Amber, our almost three-year-old goldfish, who has accompanied me through some very gloomy writing hours, yawning at just the right moments and serving as a visible reminder of this underwater world that I love so much.

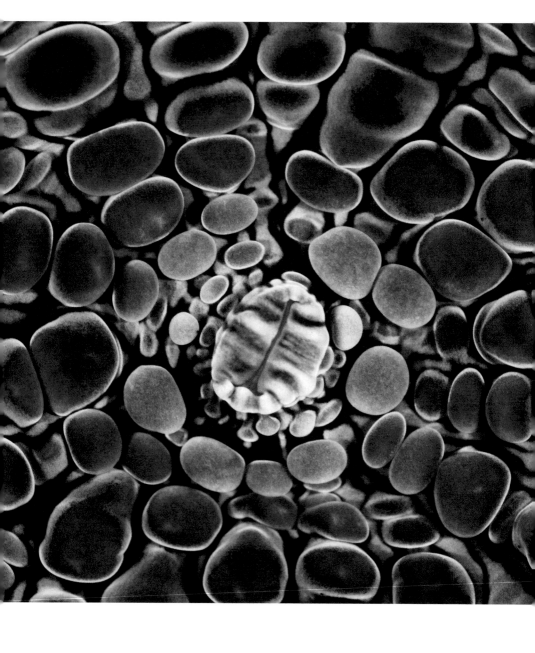

Introduction

Coral Whisperers

IB: What does it mean, coral whisperers?

LK: The point of it is to diagnose problems with corals before they're actually dead because once they're dead, it's not helpful. So we're listening to the corals, this is how they talk.

IB: So the corals are whispering?

LK: Coral whisperer means I'm whispering to the corals. But the coral is whispering back.

—Author's interview with Les Kaufman, 2017[1]

I was not even twenty when I completed my two-year mandatory military service in Israel. In an attempt to help me return to civilian life, my father purchased a weeklong scuba diving course in Sinai for me. That week of diving alongside corals in the Red Sea was transformative. Immersed in saltwater, enchanted by the kaleidoscopic array of color and form, and utterly dependent on the air in my tank, I was reminded of the connectivity among all forms of life and of the importance of returning to my own breath. Life seemed so undisturbed in the tranquility of the sea. Whatever pettiness and cruelty happened on land, it would always be met with unfaltering equanimity underwater. Or so it seemed in 1989.

FIGURE I. In 2014, this knobby cactus coral (*Mycetophyllia aliciae*) was rescued from Port-Miami, Florida, by Coral Morphologic, a Miami-based multimedia group, just before the Army Corps of Engineers began dredging operations in the area. In 2018, this particular specimen is still aquacultured in Coral Morphologic's lab. Courtesy of Coral Morphologic.

Fast-forward to 2014. My explorations of wildlife management across the wild-captive divide led me to visit a couple of coral nurseries in Culebra, a small island off the shores of Puerto Rico. As a new mother and a mid-career academic in Buffalo, New York, I had forgotten all about corals and oceans. When I went underwater to observe the Caribbean reefs, I couldn't believe my eyes. "Where are the reefs?" I asked my coral scientist guide, Edwin Hernández-Delgado, who was very proud of his restoration project and seemed slightly offended by my question. Granted, this was the Caribbean and not the Red Sea. Still, what I saw deeply saddened me: the vibrant and diverse colors of the underwater forests that I remembered from my youth had been replaced here by yellowish coral, thinly scattered over an otherwise barren-looking surface. I decided to try and understand what had happened to corals over this twenty-five-year period.

Since I am not a biologist, and because I could not easily figure out how to speak to the corals themselves, it made the most sense to speak to the people who care for and seem to know them well: coral scientists. This approach also resonated with my broader interest in science as a particular way of producing knowledge and, specifically, in biology. So, a few days after our dive, I asked Hernández-Delgado to put me in touch with other coral scientists. This minor request ultimately led to a massive research undertaking that stretched across multiple continents and disciplines. More than four years, hundreds of interview hours, and thousands of miles later, I am now clear about one thing: the relatively short period between 2014 and 2017 during which I conducted my research for this book has been transformative both for reef-building corals and for the scientists who study them. As it happened, the duration of my research coincided with what has come to be known as the third global coral bleaching event on record, a designation I was unfamiliar with until my dive in Culebra. Having started in June 2014, this bleaching event—which marked the largest and most severe coral death in human history—was declared over in May 2017,[2] precisely as I was writing this book's conclusion.

Coral Whisperers thus captures a key moment in the history of coral reef science, and of environmental conservation more broadly. It documents the physical, intellectual, and emotional plight of coral scientists and their painstaking deliberations as they struggle to understand and save corals from what many of them have come to see as the corals' inevitable catastrophic future on a polluted and rapidly warming planet. Drawing on in-depth interviews with one hundred coral scientists and managers, the book shows how, in the face of this great acceleration[3] in coral decline, coral experts are becoming the vanguard of conservation in the Anthropocene.

We are in the thick of contemporary coral science here and can feel its urgency: the experts, who are witnessing massive coral death around the planet, both grieve for this death and must simultaneously narrate it. In this role, they oscillate between positions of despair and hope. The despair-hope divide, I will argue, tracks an emergent tension between traditional approaches to conservation and newer, more interventionist, perspectives. But despite their vehement disagreements about the right thing to do, coral scientists all share a deep appreciation and intimacy, some would say love, toward corals. Through their passionate narratives, corals emerge as both a sign and a measure of the imminent catastrophe facing life on earth. At the same time, corals can also show us the way out of this apocalyptic mode and beyond the hope-despair pendulum—both so characteristic of conservation in the Anthropocene—toward what may become a more relational, or "coralated," world.

DESPAIR AND HOPE IN THE ANTHROPOCENE

The phenomenon that most struck me after interviewing coral scientists for a couple of years is that they are increasingly caught between two extreme emotional states: despair and hope. Driving one extreme are the catastrophic predictions by some scientists of a mass death of reef-building corals by midcentury, which highlight climate change and ocean acidification as the last nails in the coffin of human assaults on coral ecosystems.[4] In this despondent narrative, reef-building corals are doomed, and nothing short of an abrupt (and many would say unlikely) shift in how we use fossil fuels will save them. As the central proponent of this worldview, Australian coral scientist Ove Hoegh-Guldberg told me in an interview: "The Titanic is sinking and all we are doing is rearranging the chairs to get a better view."[5] The widely reported and minutely recorded coral bleaching event that hit the Great Barrier Reef between 2015 and 2017—wherein huge tracts of coral colonies lost their symbiotic algae and embarked on a quick and whitened path toward death—validated this pessimistic side of the pendulum, which comes replete with daunting algorithms, images, and maps. "This has changed the Great Barrier Reef forever," one scientist lamented.[6] The world's largest living structure has become the world's largest dying structure.[7]

Driving the other extreme are the scientists' narratives of hope for coral futures to come. "Give me the dying corals that you have given up on and I will restore them," pleaded a coral restoration scientist to a large and visibly uncomfortable audience at an evening plenary at the International Coral Reef Symposium held in Hawai'i in June 2016.[8] Trendy terms—like *bright spots*, which highlights locations where humans have been able to strengthen coral resilience and reverse the trend for coral death,[9] and *assisted evolution*, the selective captive breeding of "super corals" for transplantation in order to strengthen depleted coral colonies[10]—have been coined to represent actions that humans can take in the face of the growing threats to coral life. "From despair to repair," as marine ecologist Jeremy Jackson put it.[11] Under the hashtag #OceanOptimism, marine biologist Nancy Knowlton, joined by a growing number of scientists, managers, and activists, has been circulating good news about the oceans.[12]

Knowlton and Jackson fell in love with the reefs of the Caribbean and, simultaneously, with each other, some forty years ago. For decades, they were referred to by their students and colleagues as "Doctors Doom and Gloom"—until they called for a move "Beyond the Obituaries"[13] and toward ocean optimism. "We don't want people going into a catatonic state of sitting in the corner and moaning because all is lost," marine ecologist Peter Sale reflected on this turn to a more hopeful worldview.[14] Indeed, many coral scientists and managers are shifting away from catastrophe-centered narratives to focus on collaboration and hope, highlighting that "feeling hopeful enhances our capacity to take meaningful action."[15]

Both stances toward the future of coral life on earth—namely, the hope and despair ends of the pendulum—are invested with considerable emotional intensity. What I found to be most striking, however, is not so much the fervor of these stances as the disconnect—the alienation even—between the coral scientists who hold them. For instance, at the International Coral Reef Symposium in Hawai'i, I witnessed cynicism and contempt, especially from the self-appointed "realists" toward the restoration people. "Restoration is crap," one Australian coral ecologist told me. While others may not have been so blunt, the tensions surfaced clearly and produced several awkward moments at the symposium, despite the attempt by the organizers to present a united front for the sake of saving corals.

In my work studying coral scientists, I am interested in how the tensions between traditional coral conservation and newer approaches like coral restoration play out, and in what they represent. In a nutshell, I would offer that such tensions go deeper than personal and professional disagreements. They signify the ongoing, and arguably intensifying, rift between those conservation scientists who still assume that it is possible to use traditional conservation tools (chiefly the removal of adverse local human impacts) to allow natural systems to restore themselves to a prior state, and others who hold that even with the deleterious impacts removed, natural systems will not return to a prior state because their environment has fundamentally altered. These latter conservation scientists are prepared to consider active interventions of one type or another.[16] Within the interventionist approach, some go so far as to depict nature as a garden that can and must be intensely managed in order to save it.[17] Rather than perceiving humans as "screwing it up," as many traditional conservationists would have it, such radical interventionists see hope in human-nonhuman networks and collaborations and seek to foster such connections.[18]

These ideological distinctions translate into different modes of environmental management: one highlights the importance of preserving and protecting existing habitats, for example through the establishment and maintenance of marine protection areas such as the Great Barrier Reef Marine Park; the other highlights possibilities of active restoration as well as the capacity of technological advances to create novel and better-adjusted organisms and species through strategies such as assisted evolution and the design of super corals. The differences between the two ideologies and their respective modes of operation are only magnified by the grave climate predictions already afoot.

Over time, I began to notice that the bifurcated approaches toward nature within the community of coral scientists often align with gender and geography. Specifically, female scientists, many of them young and with diverse backgrounds, have taken the lead in promoting narratives of hope and models for assisted evolution. This new leadership is noteworthy in a discipline traditionally dominated by older white men.[19] As for geography, the world of coral scientists is differentiated among regions. For example, the Australians have been historically and culturally more inclined toward traditional preservation models such as securing marine protection areas, while their colleagues in the Caribbean have focused more strongly on restoration and other intensive management strategies. Currently, however, the Australians are reassessing their approach, as corals have suffered a serious death toll in their region, too.[20] In a 2017 article published in the magazine *Nature* and entitled "Coral Reefs in the Anthropocene," thirteen prominent coral scientists, many of them based in Australia, announced that "it is no longer possible to restore coral reefs to their past configurations." They argued that "instead of attempting to maintain or restore historical baseline assemblages, the governance and management of coral reefs will need to adapt continuously to the new conditions of the coming centuries."[21]

Such "new conditions of the coming centuries" have also been referred to, controversially,[22] as the Anthropocene. The Anthropocene, or the Age of Man, is an unofficial geological term coined by Paul Crutzen in 2000 and defined as the time in which the collective activities of humans have substantially altered the earth's surface, atmosphere, oceans, and nutrient cycles.[23] Corals are on the front lines of this purportedly new era, and the scientists who study them have likewise found themselves on the front lines of conservation science. There is therefore much to glean from their experiences that will be relevant to conservation scientists working with many other species and ecosystems in the near future.

The existential crisis of coral scientists and their bifurcated response to this crisis is the above-water story that I tell in this book. Underwater, another, much less bifurcated, story emerges: that of the corals themselves.

CORAL BIOLOGY ON THE FLY

Coral is a generic name for more than 2,500 species of colonial invertebrates. Relatives of sea anemones and jellyfish, corals are different in that, like trees, they are fixed to one spot for the majority of their lifetime. But coral species are also different from one another. Stony and soft corals; deep, mesophotic, and shallow corals—these are just a few distinctions that divide the various forms of life organized under the broad title *coral*. For the most part, this book focuses on reef-building corals—corals who live in the shallow waters of tropical areas and rely intensely on a symbiotic relationship between the animal polyp, microscopic algae called *Symbiodinium*,[24] and a wide range of bacteria and viruses.

Most reef-building coral species are colonial. They are comprised of multiple polyps, who typically measure only a few millimeters long and share the same genetic makeup as hundreds or even thousands of their fellow polyps who make up the coral colony. A polyp includes a stomach (gastrovascular cavity), which opens into a central mouth surrounded by a set of tentacles.[25] Hard corals also have a skeleton, or corallite, at their base, into which the polyp retreats during the day. Because of their morphological similarity to higher-order plants, in his fourteen-volume *Cyclopedia of Natural History* (1552), Edward Wotton named corals *zoophytes* (literally, plant-animals).[26] Although this term was eventually abandoned, K. Brandt's discovery in 1881 of photosynthetic algae inside the tissues of many of these tiny animals confirmed their vegetal nature.[27] These algae, most abundant in the tentacles and oral disc of the cnidarians (a large taxonomic group of over ten thousand animal species, including corals, jellyfish, and hydra, that use specialized cells for capturing prey), are still commonly called *zooxanthellae*. However, this term is now understood to apply specifically to the genus *Symbiodinium*: organisms who convert atmospheric carbon dioxide into organic carbon compounds such as carbohydrates and produce oxygen inside the animal, who is often referred to as the "host."[28]

The mutual benefits of cooperation to both animal and algae are at the heart of the symbiotic relationship that characterizes reef-building corals. Reef-building corals are phototrophic organisms: the *Symbiodinium* algae live inside the polyps and provide them with nutrients derived from photosynthesis. This relationship with the algae helps "speed up the process by which corals build their stony skeleton, putting down layer after layer of calcium carbonate, a form of limestone. It is this process of calcification that physically constructs the bedrock of the coral reef."[29] Without their symbionts, reef-building corals would not have such high rates of growth and thus would not be able to form reefs as we know them.

Corals are both autotrophs, able to derive energy from the sun, and heterotrophs, able to absorb nutrients from the environment by ingesting microorganisms such as bacteria and zooplankton who drift through the oceans. A recent study showed that during bleaching events, certain coral species were able to maintain and restore energy reserves by increasing their feeding rates.[30] As for the algal symbionts, they are genetically diverse, comprising nine evolutionary lineages ("clades") that share a common ancestor from approximately fifty million years ago. Alongside *Symbiodinium*, the coral assemblage includes a vast array of microbial symbionts such as bacteria and viruses.[31] This entire symbiotic assemblage is referred to by coral scientists as a *holobiont*.

When the ocean water becomes too warm, the *Symbiodinium* algae living in tropical coral tissues are expelled, causing the coral to turn completely white. This process is called *bleaching* (figure 2). It is still unclear who initiates the expulsion, the animal or the algae, but the evidence leans toward the latter.[32] The algae also apparently "choose" which corals to "infect" in the first place—

FIGURE 2. A scientist surveys the bleaching at Heron Island in the southern Great Barrier Reef, February 2016. Courtesy of The Ocean Agency / XL Catlin Seaview Survey.

and I use scare quotes here because many scientists would contest this implied agency by the algae and see it as problematically anthropomorphic.[33] "It happens once in a very early stage, just after the larva settles, and then the window of opportunity closes, nobody [else] can get in—the coral will just eat the latecomers up," coral geneticist Mikhail Matz told me, not shying away from agency and anthropomorphism himself. "There is only one winner when you infect the coral," he clarified. "It's like a little bit of natural selection within that particular tiny little coral polyp."[34] Marine biologist Mary Alice Coffroth noted that "it can take up to four years until the final *Symbiodinium* type is established, although it probably gets in within the first three months."[35]

A bleached coral is not dead, yet. Corals can survive a bleaching event, but such an event puts them under increased stress and they are therefore more likely to die.[36] Research has shown that when recovering from bleaching, coral species and even colonies cooperate with their original strain of symbionts.[37] Since the 1980s, episodes of coral bleaching and mortality, due primarily to climate-induced ocean warming, have occurred almost annually in one or more of the world's tropical or subtropical seas.[38]

Alongside such local episodes of bleaching, in 1998 the U.S. National Oceanic and Atmospheric Administration (NOAA) announced a new type of event: global bleaching, defined as mass bleaching of at least one hundred square kilometers in all three ocean basins.[39] Triggered by the El Niño of that year, the first major global bleaching event was declared when a huge underwater heat wave killed 16 percent of the corals on reefs around the world's oceans. The second global bleaching event was caused by the El Niño of 2010. In 2015, NOAA announced the third global bleaching event, the longest on record, which impacted reefs around the world between June 2014 and May 2017. It hit the Great Barrier Reef particularly hard, with 93 percent of the surveyed reefs bleaching between 2015 and 2017. During that time, some locations in Hawai'i saw 75 percent of their corals bleached.[40] As for the future, climate models predict that most of the world's reef-building corals will face annual temperature extremes before the end of this century, with some experiencing such conditions starting as early as 2030. Scientists are currently documenting how this grim forecast unfolds from moment to moment, producing a painfully detailed testimony about the process of corals becoming extinct.

Yet many members of the public find it hard to relate to coral death on such grand scales. For me, the face of bleaching is the pale look of Edwin Hernández-Delgado—the scientist who introduced me to his coral nurseries during our 2015 dive in Culebra—when he noticed that the corals he had transplanted were turning white. The correlation between the coral's whitening polyps and the scientist's whitened face is, poignantly, yet another sign of our "coralated" materialities.

CORAL ECOSYSTEMS UNDER THREAT

Humans have only recently come to recognize that coral reefs are living beings, not simply rocks, and that their existence is under threat by our capitalist modes of consumption. In "From Threatening to Threatened: How Coral Reefs Became Fragile," historian Alistair Sponsel observes how human perceptions of corals have changed through time—from seeing them as hazardous impediments to navigation to seeing them as fragile living organisms who are threatened by humans and who should thus be protected and saved.[41] In biopolitical terms,[42] corals have transformed from killable to grievable creatures.[43]

Today, tropical coral reefs are considered among the most diverse marine ecosystems on earth, providing shelter to myriad species, including four thousand species of fish and another one to eight million still undiscovered species of organisms living in and around reefs.[44] Coral scientists like to emphasize that although reefs represent less than 0.1 percent of the world's ocean floor, they help support approximately 25 percent of all marine species, which is why they are commonly referred to as the rainforests of the ocean.[45] As a result of their precarious state, the livelihoods of five hundred million people and an income worth over $30 billion are at risk.[46] Coral reefs also provide buffers that protect shorelines against waves, storms, and floods. The absence of these natural barriers will increase the damage to coastal communities from wave action and violent storms.[47]

Coral scientists warn that at present, corals are facing multiple stresses caused by pollution, overfishing, and, increasingly, global warming and ocean acidification. The scientists further contend that corals act as an early warning system, their alarming status reflecting the deteriorating health of the oceans. Coral reefs are thus referred to as canaries in the coal mine. If they radically decrease or even disappear, this analogy implies, other marine life will soon follow, and human life will be severely affected.[48]

Reef-building corals have disappeared several times before. The most recent event happened at the end of the Cretaceous era, approximately sixty-five million years ago.[49] After each of these catastrophes, it has taken roughly ten million years for reef building to recommence. Although in the 1970s most reefs showed slightly positive growth rates, this trend reversed course shortly after, and coral populations have been declining ever since. What we have seen over the past few decades, and can expect to see in the foreseeable future, is an exceptionally rapid and global death of reef-building corals. The coral thus emerges as the less-like-us cousin of the polar bear—the new "poster child" of climate change.

But while being in the spotlight can be a good thing for coral protection, it may also carry a price. For under climate change's all-consuming shadow, it is easy to lose sight of other conservation threats and potential courses of action—and, indeed, to overlook the corals themselves as diverse and complex creatures. In this sense, embracing the reality of climate change has, ironically, hindered coral conservation.[50] In the words of Jeremy Jackson: "The fascinating thing about climate change is that it's an excuse for doing nothing."[51] Alongside documenting the recent shift of conservation science toward a focus on global warming, this book also records the attempts by many coral scientists to mitigate its effects through targeted local action and by crafting alternative balancing schemes. Historian Dipesh Chakrabarty reminds us, accordingly, that "the climate change problem is not a problem to be studied in isolation from the general complex of ecological problems that humans now face on various scales—from the local to the planetary—creating new conflicts and exacerbating old ones between and inside nations."[52]

THINKING WITH CORALS

Corals are "good to think with."[53] The symbiotic algae-animal relationships at the core of their precarious existence reveal that, more than a single unified entity, corals are, in fact, "coralations"—bundles of constantly changing assemblages that shape and reshape their ways of being in the world. Beyond the symbiotic underpinnings of their microscalar existence, coralations also occur at the level of the coral colony and ecosystem as well as at the intersections of culture, science, and law.

Corals confuse and destabilize our categories: they are a cross between animal, plant, rock, microbe, and ecosystem; we sentimentalize them because of their beauty, despite the fact that they don't have a face or a clear sex[54] and so we can't easily anthropomorphize them; and while they live in the ocean, which constitutes the majority of the earth but which we know so little about, they also constitute some of our terrestrial mountains and buildings.[55] Reef-building corals are animals, yet they photosynthesize; they make massive stony structures that can be seen from space, but they are tiny and, some claim, fragile creatures; they are sessile, yet travel long distances in their larval stage; and each has a mysterious symbiotic relationship with a particular strain of algae—who, under certain conditions, disembark from the coral cells, leaving them bleached and depleted. Individual coral polyps in a colony may differ in morphology and genetics, and some may be fusions of two or more genotypes. For the most part, however, polyps who belong to one colony have the same genetic composition—what scientists refer to as "ramets." Coral colonies are interconnected by living tissue. Finally, unlike most animals known to science, they don't really age: given the right conditions, corals can live forever.

Throughout history, corals have inspired indigenous cultures, poetry, and art. The Kumulipo ("Beginning-in-deep-darkness") is the sacred creation chant of a family of Hawaiian ruling chiefs. Composed and transmitted entirely in the oral tradition, its two thousand lines provide an extended genealogy detailing the family's divine origin and tracing its history from the beginning of the world.[56] The Kumulipo opens with the coral as the first organism in the Hawaiian universe. Corals are the beginning of life, the most ancient ancestors of all living things.

Corals have also fascinated great intellectuals such as Karl Marx, who mentions them as prime examples of the relationship between the individual and the community.[57] Charles Darwin's first monograph in 1842 was entitled *The Structure and Distribution of Coral Reefs, Being the first part of the geology of the voyage of the Beagle, under the command of Capt. Fitzroy, R. N. during the years 1832 to 1836*.[58] More recently, the coral made a prominent appearance in revered evolutionary biologist and historian of science Stephen Jay Gould's last book, *The Structure of Evolutionary Theory*.[59] There, Gould used an image of a coral to represent the basic ideas of Darwinian theory.

Ironically, many scientists who have studied corals have come to challenge the traditional Darwinian principles of evolution, highlighting the centrality of the symbiotic relationship and the importance of understanding the coral as a holobiont (again, a composite of coral animal and a diverse set of microbes, including algae). These coral-spawned realizations have brought about a substantial paradigm shift in the field of biology, which was until recently dominated by neo-Darwinian theories of origin and natural selection.[60] It is thus not very surprising that scientists who study corals also promote a rhizomatic outlook on the world. The "rhizome" is a concept developed by French philosophers Gilles Deleuze and Felix Guattari in the 1970s to highlight ways of thinking that are multiple and nonhierarchical, as opposed to "arborescent" (tree-like and hierarchic) knowledge that works with dualistic categories and binary choices.[61] As Darwin himself acknowledged in his notebook, "The tree of life should perhaps be called the coral of life,"[62] implying that his own view was much less "Darwinian" than it was later interpreted to be.

This is just a sliver of what happens when one starts thinking with reef-building, tropical corals, commonly referred to as stony corals or scleractinians. Add to this the existence of soft corals (gorgonians, or octocorals), who have very different skeleton composition and symbiotic relationships, and who possess eight tentacles (hence *octo-*) instead of the multiple sixes of the scleractinians, and you get a mind-blowing diversity that raises a set of important questions (see figure 3). For example, scientists have observed that soft corals don't bleach as much as their stony relatives. They are now asking what this means for rapidly transforming coral ecosystems in polluted, overfished, and warming oceans. Who are these corals of the Anthropocene, and what can we learn from them about the corals of the future?

And here's another coralated question to consider: only in the past few decades have marine scientists discovered that at least half of all coral species

FIGURE 3. Ángela Martinez Quintana was born in Galicia, Spain, and is currently studying for a Ph.D. at the University at Buffalo, SUNY. In her words, "This octocoral is called *Plexaurella dichotoma*. I am studying this coral in the U.S. Virgin Islands. I took this picture while I was doing a census of octocorals, a major component of many Caribbean reefs. These corals are typically fifty to seventy centimeters high; they also have *Symbiodinium* and bleach at high temperatures. But unlike many stony corals, many octocorals have survived the bleaching, which is something we are naturally very interested in" (e-mail communication, June 17, 2017). Photo by Ángela Martinez Quintana, July 2016.

in fact live in the dark and cold abyss of the deep ocean. Unlike the corals near the ocean's surface, deep-sea corals do not photosynthesize or bleach.[63] But in addition to warming oceans and pollution, myriad other threats assail these unique ecosystems, such as bottom trawling, deep-sea mining, and ocean acidification—the ongoing decrease in the pH of the oceans caused by the absorption of carbon dioxide from the atmosphere. Because of the invisibility of deep-sea corals to human eyes and the relative paucity and unenforceability of international legal regimes that pertain to their habitats in the high seas, these corals are underprotected by law and marginalized within the field of coral science when compared to their shallower kin. Indeed, deep-sea scientists often complain that cold-water coral ecosystems are more alien to humans and less explored than the landscapes of Mars.[64] In this sense, immersing ourselves in the multiplicity of coral life allows us to step back and recognize the many assumptions, including legal ones, that underlie our understandings and our regulation of life and death—both in our environment and, eventually, in ourselves.

But perhaps I am overstating the importance of thinking with corals. The emotional layers in the work of coral scientists are arguably just as meaningful as their analysis, if not more so.

SCIENTISTS ON THE BRINK: A NOTE ON METHOD AND EMOTIONS

The account in this book is both eclectic and entangled. It was almost as if the corals themselves—with their fragmented, reticulate, and unwilling-to-be-classified nature—whispered their resistance to me and left me spellbound. Inspired by Anna Tsing's patchy ponderings about mushrooms[65] and by Paul Rabinow's direct presentation of scientific narratives,[66] this book focuses on coral scientists attempting to understand their world both individually and as a community. It is organized around their oscillation between hope and despair, with climate change playing an intensifying role for both sides of the pendulum.

Such swings from hope to despair and back again are certainly not the sole province of contemporary coral scientists, although they are quite acute in this context. Myriad recent books about conservation in general, and about climate change in particular, embrace one emotional extreme or the other. Despair titles include philosopher Dale Jamieson's *Reason in a Dark Time: Why the Struggle against Climate Change Failed—and What It Means for Our Future* (2014),[67] historian Joshua Howe's *Behind the Curve: Science and the Politics of Global Warming* (2014),[68] climatologist James Hansen's *Storms of My Grandchildren: The Truth about the Coming Climate Catastrophe and Our Last Chance to Save Humanity* (2010),[69] political scientist Stephen Meyer's *The End of the Wild* (2006),[70] journalist Elizabeth Kolbert's *The Sixth Extinction: An Unnatural History* (2014),[71] and environmentalist Bill McKibben's early iteration of this mood, *The End of Nature* (1989).[72]

Hope, on the other hand, is the central theme in countless other titles, including psychologist Mary Pipher's *The Green Boat: Reviving Ourselves in*

Our Capsized Culture (2013),[73] philosopher Jonathan Lear's *Radical Hope: Ethics in the Face of Cultural Devastation* (2006),[74] literary scholar Teresa Shewry's *Hope at Sea: Possible Ecologies in Oceanic Literature* (2015),[75] environmental activist Joanna Macy and psychologist Chris Johnstone's *Active Hope: How to Face the Mess We're in without Going Crazy*,[76] conservation scientist Andrew Balmford's *Wild Hope: On the Front Lines of Conservation Success* (2012),[77] and *Climate of Hope: How Cities, Businesses and Citizens Can Save the Planet* (2017)[78] by former New York City mayor Michael Bloomberg and former executive director of the Sierra Club Carl Pope.

Increasingly, scholars are highlighting the need to balance between the despair-hope extremes. "How might it be possible to move environmentalism beyond the stereotypical narrative of the decline of nature without turning it into progress boosterism?" asks cultural scholar Ursula Heise.[79] Donna Haraway argues, similarly, that "there is a fine line between acknowledging the extent and seriousness of the troubles and succumbing to abstract futurism and its affects of sublime despair and its politics of sublime indifference."[80] Finally, Teresa Shewry calls for an understanding of hope beyond hope: "Hope is not well understood simply as the loss of awareness of reality . . . or as entirely distinct from sorrow, despair, and other experiences," she writes. Instead, "hope is offered through attunements to present world struggles to enliven environmental relationships that have been lost or damaged. As such, it is never alone; it always exists intimately with experiences such as sorrow."[81] Although on the surface each is experienced as exclusive of the other, when examined more deeply one can begin to see the ways in which the hope-despair narratives by the coral scientists in this book are interconnected and even contemporaneous. This recognition will emerge within the chapters and will surface again in the book's conclusion.

My decision to write a book about coral scientists and their oscillation between despair and hope was far from obvious, and not without its perils. I am keenly aware that "the primary and sometimes exclusive focus on science . . . is what drives the failure of climate change advocacy," as Joshua Howe put it.[82] His critique is twofold, as he highlights science's fundamental inability to state anything with 100 percent certainty as well as its problematic conviction that the solution to uncertainty is yet more scientific research. Another problem with focusing on scientists emerges at the emotional level. Dale Jamieson warns in this context that "ignorance of science can give rise to excessive respect, which can quickly turn to disillusionment when science does not deliver the goods or when scientists turn out to be as petty and selfish as the rest of us."[83] Relatedly, geographer Jessica Barnes and social ecologist Michael Dove write about the "unexpected and resolute mistrust of this science in many segments of the public, especially in the United States."[84]

Yet rather than seeking to move beyond the perspective of scientists, this book fully embraces it. Indeed, *Coral Whisperers* is an attempt to see corals as coral scientists see them, to understand the world as they understand it. The phrase "on the brink" intends to capture the sense of urgency and crisis that has prevailed throughout my discussions with these scientists as they attempt to save the corals from imminent death and themselves from imminent obsolescence. Mary Pipher argues that "there is an enormous gap between what scientists know and what most people know."[85] This is especially true with regard to oceans, which many of us don't experience regularly and don't know much about.

But instead of focusing only on what scientists know, argue, and do in their professional life, *Coral Whisperers* depicts coral scientists as people and situates them within a dynamic community. I have therefore not shied away from relaying the fierce debates that split this community, nor from pointing to the care and anguish that unites them. Rather than turning to narratives of non-scientists, as Howe calls for, and while being careful not to glorify the scientific enterprise, this book works to dispel the myth of a homogeneous and objective science, at the same time granting scientists a voice by accounting for their real-world experiences and deliberations.

In dealing with the volume of interviews and observations I've gathered over the course of several years, my intuition has been to let the scientists speak in a less mediated way. This means that the book contains quite a bit of science—and some rather long quotes. Coral scientists from around the world emerge here not only as harbingers of doom and messengers of hope, but, perhaps more importantly, as humans facing an existential crisis. Their task is tough: they must both witness their beloved corals dying out and at the same time narrate this death to the world. They must be careful not to sound too gloomy, lest their warnings be labeled alarmist, nor too hopeful, lest they be dismissed as Pollyannaish or even as denialists. Finally, all this is happening in the face of what many coral scientists experience as disinterest in coral death and its causes on the part of the general public and certain prominent politicians.

In this dramatically changing world, under the looming shadow of climate change, the role of the coral scientist is also undergoing dramatic transforma-

FIGURE 4. Marine biologist Lorenzo Bramanti of the French National Centre for Scientific Research (CNRS-LECOB) uses a syringe to collect organic and inorganic matter trapped in a canopy of corals and also to collect their larvae, which he then keeps in an aquarium for experiments. Bramanti told me about his relationship with the red coral: "There are so many reasons I feel connected to the red coral of the Mediterranean. I'll start with magic. According to Greek mythology, when Perseus killed the Medusa, the blood spilling from her cut head splashed on some marine algae that became red, giving origin to the red coral, referred to as *Corallium rubrum* by scientists. Because of its hard and red skeleton, the Mediterranean red coral has been harvested for centuries and is so precious that it is often referred to as 'red gold.' As a result, nowadays the red coral can only be found below a depth of forty meters. When I headed to the university to become a marine biologist, I realized that I wanted to work on this one precious coral that required specialized diving skills and that was surrounded by an aura of myth. I spent my bachelor's, Ph.D., and two postdocs—more than twenty years—working on the ecology, biology, reproduction, conservation, and management of the red coral. It is summertime now, and I am awaiting the full moon with anticipation—this is when the red coral spawns. Just a few days ago, I found larvae in the aquarium where I put the mother colonies. These are the babies of the corals that I collected from the sea. I feel like they are my babies" (e-mail communication, June 16, 2017). Photo by Francisco Romero, July 2017, Cap de Creus, Cataluña, Spain.

tion. In the not-so-distant past, scientists focused on nuanced studies of their subject species; now they must become spokespersons for their corals if they are to save them. Although many of my interviewees felt uneasy speaking about their emotional bond with their coral subjects and were also quite wary of taking any sort of political stance—and they are not different in this regard from scientists in other disciplines[86]—I also encountered a growing cohort of scientists who were prepared to do so, and boldly (see, e.g., figure 4). I soon noticed that many of the latter scientists were women, and have come to see

FIGURE 5. Jenna Budke and Katia Chika-Suye of the University of Hawaiʻi at Hilo replace data loggers on a coral reef at the Waiʻopae tide pools, Hawaiʻi. Photo by author, June 28, 2016.

these individuals as the "Jane Goodalls" of coral science.[87] Although an outsider to this community, I, too, have been struck by the tragedy of coral demise and humbled by the persistence, dedication, and love for corals that I have witnessed on the part of so many of the scientists I interviewed.

I'd like to share just one example of this dedication from my fieldwork in Hawaiʻi. When we came to the surface during a work dive on the Big Island, local Ph.D. student Katia Chika-Suye spontaneously expressed her appreciation for the blue rice coral (*Montipora flabellata*), which is endemic to these islands. "Isn't it so beautiful?" she asked me, not awaiting my answer as she dove right back down, with me lagging behind. I had joined Chika-Suye and her colleague Jenna Budke for a morning dive at the Waiʻopae tide pools in 2016. We were there to ensure that the temperature-monitoring equipment was still operative and to collect samples of a few coral species (figure 5). By breaking several of these corals off from their colony with a chisel and hammer and observing them under a miscroscope, Chika-Suye and Budke would find out whether the corals were getting ready to spawn, in which case they would collect their eggs. Just when I thought I would lose my fingers and toes from being submerged in the cold water for what seemed like forever, my belly grumbling from an overdose of saltwater inhaled in the high and rough tide, we finally gathered all of the instruments and samples and headed back to shore.

I was grateful for the sun's warmth and for not having to struggle for every breath. Under the microscope, the samples we had collected provided an intricately beautiful scene: what had looked no different than dead rock revealed itself as very lively and exquisite tentacles connected to a body containing tiny and perfectly shaped eggs. When we finished studying these creatures, Budke headed straight back to the water. I was too exhausted to join her, but when I asked about it later, she explained that she had returned the coral fragments back to the ocean, with the hope that they would reattach and survive.[88]

Many months later, I met with a different group of coral scientists, this time at the Boston University marine lab. In that instance, while the younger female graduate students shied away from making statements about intimacy with their coral subjects, the lab's senior male professors strongly emphasized the depth of this relationship. One of them, Les Kaufman, stressed that to be the person who knows a coral species better than anyone else in the entire world carries with it an immense responsibility to serve as the spokesperson for this coral, who cannot speak for herself.[89] "Now this is intimacy!" Kaufman exclaimed.[90] Everyone in the room was in complete agreement.

When I reflect on the many hours that coral scientists have taken from their work in order to speak to me, I realize that this is precisely what they were doing: serving as spokespersons for their corals and hoping that the message of urgency, love, and grief would travel through me to a larger public. "Maybe, in your book, you can make people understand what an extraordinary animal the coral is and how horrible a world without them would be," one scientist confided in me after our formal interview was over.[91] The term *coral whisperer* thus takes on an additional meaning, now including the whisper's intended audience. The whisper of the scientists, the whisper of the corals, must both be magnified so that they may be heard by this audience. My book undertakes an atypical journey into the world of those who whisper to corals and to whom corals may whisper back. It is my hope that it will serve to amplify these whispers into the piercing cry that they deserve.

THE STRUCTURE OF *CORAL WHISPERERS*

Broadly stated, this book is a chronological documentation of how scientists have come to gradually realize and respond to the magnitude of threat to their beloved coral species and ecosystems—from their initial insistence to hold on to traditional conservation methods and to projects of monitoring (chapter 2), to their growing reliance on restoration (chapter 3), through their collaborations with legal actors and regulatory projects (chapter 4), and, finally, to their recent turn to coral genomics and assisted evolution (chapter 5).

Beyond this chronological account, however, *Coral Whisperers* is structured as a pendulum swinging between hope and despair. Chapter 1 explains the workings of this pendulum and the stakes involved. Relying considerably on my observations at the International Coral Reef Symposium in Hawai'i in June 2016, this chapter relates the intellectual and emotional debates among scientists about how best to manage the impending catastrophe facing coral reefs. This chapter is, in many ways, a snapshot of the entire book, depicting the oscillation between hope and despair and telling the stories of particular scientists who have experienced this oscillation firsthand.

Next, chapter 2 documents the despair side of the pendulum as it contemplates the existing modes and technologies for recording coral bleaching and death. Here, the trajectory typically tends toward devastation and gloom, as the numbers are depressing at best. Much of the chapter focuses on the third global bleaching event at the Great Barrier Reef, documenting how scientists have both recorded and narrated this event to themselves and to the general public. I examine the role of monitoring in particular, considering whether enhancing scientific knowledge about corals by monitoring them is an act of hope, in that it supports conservation action, or one of despair, in that it stifles such action and masks the resulting inaction with more and more monitoring. Finally, the chapter shows that even in the world of numbers and maps, "bright spots" and optimistic indexes still rear their more hopeful heads.

In chapter 3, the pendulum swings again, this time to document acts of hope by coral restoration scientists. Drawing on my visits to five coral nurseries—Culebra in the Caribbean, southern Israel, Honolulu and Coconut Island in Hawai'i, and the Florida Aquarium—this chapter explores the scientific, cultural, and emotional challenges facing restoration efforts and the criticisms directed toward these efforts from within the coral science community. A marginalized field only a few years ago, coral restoration is becoming increasingly important to a growing number of scientists and managers, who see it as a way to resist death by engaging in coral propagation. The coral nursery assumes center stage in this process and emerges as part of the broader attempts by restoration scientists to establish coral restoration as a science that is aligned with silviculture and restoration ecology. The coral nursery is finally compared to the coral farm, where corals are propagated for the mariculture industry. Throughout, the chapter discusses the often fraught relationship between biologists, aquarists, and hobbyists, pointing to the importance of enhancing collaborations between these groups for coral survival.

In chapter 4, the pendulum swings back to despair. The chapter documents the focus of contemporary legal regimes, and the U.S. Endangered Species Act in particular, on threat and endangerment, exposing how ill-suited existing law is for dealing with the weirdness of corals and with the sheer scale of their "super wicked problem": climate change.[92] Here, federal administrators provide insight into the processes through which listing and delisting decisions are made, their deliberations about coral classification engendering a fresh perspective on the coralations between law and science. While the relationship between scientists and lawyers is often fraught, and both express frustrations with existing legal regimes, some also see hope on the legal horizon. Two such hopeful instances are the emerging international regulation of climate change and the signals of possible receptiveness by U.S. courts to the assertion of constitutional rights to a healthy environment.

Finally, chapter 5 explores the hope encapsulated within the corporeality of coral life. Holobionts, hologenomes, chimeras, and reticulate evolution are just a few of the concepts deployed by coral scientists to shed light on the microscopic and macroscopic complexity of corals, highlighting the emergence of a genomic turn in coral science. The chapter delves into coral symbiosis, exploring how it challenges our definitions and understanding of the individual—in corals and beyond. It also points to debates among coral scientists about the resilience of hybrids and about the efficacy and appropriateness of novel experiments in assisted evolution. These novel designs focus on tweaking present-day corals for the sake of corals of the future, creating what some scientists have called "super corals." Coral tissues may also serve as living gene banks—repositories of DNA that freeze cellular life in time until they can be thawed in the future. Finally, forensic readings of the rings of coral skeletons provide a window into past climates, especially in tropical oceans where reading ice cores and tree rings is often not possible.

Throughout the chapter, tensions emerge between field scientists and geneticists about the right way to save corals; nonetheless, it ends on a hopeful note—with an invitation for cooperation.

In between the chapters, six extended interview segments provide a first-person perspective from the following conservation experts on their work with corals: Peter Sale, Ove Hoegh-Guldberg, Jeremy Jackson, Ken Nedimyer, Murray Roberts, and Ruth Gates. These interludes allow the coral experts, most of them scientists, to speak for themselves more directly. They also shed light on the unique bond between the scientists and these imperiled animals, who are otherwise rather alien to most humans and so the scientists' relationship to them is not as obvious as it would be toward, say, a rhino or a polar bear. A majority of the book's photographs were chosen and introduced by the scientists interviewed, providing another powerful medium through which they may may relay their relationship to particular corals and environments.

Throughout this book, a sense of urgency prevails. Coral reefs are experiencing, with growing frequency and severity, mass bleaching, disease, and mortality, as carbon dioxide and temperatures increase, water quality deteriorates, and marine fishery stocks are overexploited.[93] As when trying to save a dying loved one, coral scientists can't afford to waste time or words. It is now or never—they must get it right as they embark on what, for them, is the ultimate "emotional race against time."[94]

Corals in the Anthropocene

An Interview with Peter Sale

Peter F. Sale is a marine ecologist. He has been a faculty member at the University of Sydney, Australia; the University of New Hampshire, USA; and the University of Windsor, Canada, where he is currently professor emeritus. His research in Hawai'i, Australia, the Caribbean, and the Middle East has focused primarily on reef fish ecology and on the management of coral reefs. In his 2011 book *Our Dying Planet,* Sale draws on his personal experiences to help tell the story of the global environmental crisis to a general audience. I interviewed him over Skype on July 7, 2016. I decided to place his interview, which I have arranged and edited, as the first of a series of six highlighted interviews, because it conveys the broader issues in contemporary conservation and highlights the transformative times we are living in, which Sale refers to as the Anthropocene.

IB: What, if anything, is unique about the particular time we are living in?

PS: People don't understand that we're in a rather different world now from the world we were living in ever since human civilization started. In the world that we used to live in, up until we screwed it up, you could be reasonably certain that the way things were would persist more or less into the future. In other words, if you had a forest of a particular type, that forest would be there in one hundred years or two hundred years, unless you cut it down. Maybe there would be a forest fire that would go through and dramatically change it and it would take a while to recover, but it would eventually recuperate.

That is no longer true. The world we're in now is a world where everything is changing. The forest that I have around me here [in northern Ontario, Canada] is being forced to move north at a pace that the trees cannot match. And as a consequence, over the next couple of hundred years this forest is going to change dramatically. Trees that are no longer suited to living here are going to be doing less well. . . . In all likelihood, the forest is going to deteriorate because the trees from the south are not going to get here quickly enough to replace the trees that are being lost. A sycamore tree can spread its seeds northwards but only at a certain rate, and if we want sycamore trees in northern Ontario we may very well have to carry them here. That is a totally different world than the world we used to live in.

And it applies to coral reefs as much as everything else. Of course, most of the time we try to manage reefs by creating protected areas, and then we don't actually protect them. But if we assume that we get really good and we create protected areas and we really protect [corals], the reefs are still going to change, they're still going to degrade. Because we have unleashed a warming of the climate which is proving very difficult for them, and we have unleashed an acidification of the ocean which is proving difficult for them. As a consequence, they are going to dwindle. Our protected areas do not protect against warming or acidification.

So when I hear that somebody is building a coral nursery and is propagating fast-growing corals from little bits of coral and growing them up in the nursery and then planting them out on the reef, what I see is an activity that will maybe capture people's attention—which is good—but is actually doomed if the production of these baby corals is not going to produce corals any more capable of resisting the threats than the corals that were killed off. If they are not engaged in selective breeding and some kind of process for developing clones of corals that are going to be super-resistant to temperature, or really capable of coping with low pH, [then] they're wasting their time. And it's terrible to tell people they are wasting their time, because most of the time they don't realize it—because they haven't realized we're in this different world. And so that's one of the things I think of when I hear about coral restoration.

The other criticism of coral restoration, of course, is that the scale at which everything is being done is woefully inadequate. You can grow up a bunch of baby corals and plant them out on a reef and create an area half a hectare in size that grows for a little while, but a bleaching event can come along and do significant damage to half of the Great Barrier Reef over a period of weeks. Until we match our restoration to the scale of the damages [that occur], we're really just diddling around the edges. It's difficult to see it because many of the people engaged in restoration really believe they are doing something useful. But in all likelihood, they're not. It's like building a seawall in Florida—it's a waste of time and money. Florida is going to be submerged, and we need to recognize that. And again, some of the things that people don't understand, for instance, [is that] if we stop putting any CO_2 into the atmosphere today, the temperature will keep on warming for another fifty years. That means that sea level will continue rising for another couple of centuries because of what we've already done. We've set very slow processes in motion and [even] if we stop what we're doing, we then have to give those processes time to stop. There are a lot of people who don't understand that we have already committed ourselves to a 1.5°C warming.

IB: What would you say to people who don't care about corals dying off?

PS: Coral reefs are currently 0.1 percent of the surface of the oceans. [They are] amazingly valuable because of the enormous diversity they hold, but they're relatively rare. They're going to become a lot more rare. So we're going to go through what the evolutionary biologists refer to as a "substantial bottleneck." There is going to be an enormous loss of species simply because we will not be able to have sufficient area of living reef to retain them. Plus, species that are adapted to living in particular locations where we won't get any new reef growing are all going to go. Does it matter if the world loses a substantial fraction of its marine species? Well, it's happened five times before. It happened at the end of the Permian, there was a 95 percent loss in marine genera. That was an enormous loss of diversity and it took many tens of millions of years to build it back up again. As long as we don't lose it all, evolution has a way of ramifying and multiplying—and, eventually, biotically rich ecosystems will all come back. But they're not going to come back in human lifetimes. So we are in danger of witnessing the progressive and continuing decline of the ocean's genetic variability, the

FIGURE 6. Peter Sale chose to share with me this photo of the manini fish schooling over a Hawaiian reef. He explained: "The manini, *Acanthurus triostegus*, is the first coral reef fish I researched. It is superabundant on Hawaiian reefs, but much less common on the southern Great Barrier Reef, where I did most of my work for the next twenty years. This very widely distributed, often common, surgeonfish—found throughout the Indo-West Pacific and on the western (Pacific) coast of Central America—extends from the reefs of East Africa as far east as Hawai'i and French Polynesia and even the Galápagos. Despite its ubiquity, and perhaps because of its rarity on the reefs where I worked in Australia, I came to see the manini as an old friend. Detecting one during a dive always lifted my spirits! Widely distributed, crazy-abundant, and yet a very special fish for me" (e-mail communication, July 12, 2017). Photo © Pelika Andrade.

range of species, the complexity of food webs—a whole host of things are going to get simplified. And that is probably going to have significant implications for food sources for people because we [consume] a lot of seafood (figure 6). And how that all works out, along with our growing need for food—because there are so many of us—I think it's problematic.

IB: What is your take on genetically assisting the corals?

PS: I think we should continue to do it because we will be better off with the results than without them. Do I think that there's any way that [genetic] action is going to make it possible for us to have flourishing reefs over the next hundred years? No, that is going to be impossible. There are going to be little pockets of flourishing reefs, and they are going to be the safety net that somehow gets us through a rather nasty patch in the evolution of the earth.

The field of coral genetics has exploded so rapidly. I keep being amazed at the kinds of new techniques that are constantly being developed. I mean, it started out with a guy growing peas, and it has gotten quite a bit more advanced. And there were certainly papers at the Hawai'i conference we both attended that I didn't understand. That's one of the interesting things about that conference: the breadth of sciences being applied to coral reefs is getting a lot broader, and that's good.

IB: What does the decline of coral reefs mean for you, on a personal level?

PS: The things we're doing to coral reefs are just a signal of the things we're doing to the entire ocean. And the things we're doing to the entire ocean are mirrored by the things we're doing to the entire land surface of the planet. It's possible to get exceedingly depressed about it, and I sort of try not to do that and try to focus on the things we need to do in order to do a better job. One of the things we need to do is we've got to put ourselves back into the biosphere. We've got to recognize that we're part of the living skin of the planet, we're not somewhere on the outside, managing it or using it because we're entitled. We've got to recognize that we've got to sustain it because it sustains us. And whether we're going to be able to make that leap is a huge question given that we have an exceedingly robust machine that is now global and is pushing a consumerist philosophy that says that resources are there to be used, items are in the stores for you to buy and you should feel obligated to max out your credit cards because that's what keeps the economy going. [There are] exceedingly strong pressures on us, which we have created, that keep us from seeing ourselves as part of the living world.

IB: Many of the coral scientists I've interviewed were uncomfortable with making emotional appeals. What about you?

PS: As scientists, we've been taught all our lives that we should be objective and unemotional and [that we] mustn't let our subjective feelings influence how we interpret the data. I now believe that we should be more emotional. We should convey to people not only that there are serious problems, but also how much we care about those problems and why those problems are so important to us. Because for most people, the problems we are talking about are not that important. And if we're going to convey that these are serious problems, I think it is perfectly appropriate to convey how seriously they affect us as individuals.

IB: Could you tell me a personal story on that front?

PS: Sorry, what was that?

IB: I'm challenging you to tell me your own emotional story.

PS: Okay. Well, I'll answer it in an unemotional way, but this is how it is possible for me to become emotional. I am an ecologist. I understand that living organisms on the planet interact with each other in many ways, positive and negative, and that we are one of those species and that we interact with other species. Because I'm a professional ecologist, I also understand that some of the simple ideas that are now unfortunately very well entrenched in Western thought, specifically the idea of the balance in nature, are metaphors that the reality doesn't really uphold. I'm much more prone to the belief that if we push the ecological system hard enough, it can go seriously wrong in the sense that it will change in ways we don't anticipate and probably in ways that will be detrimental to us. I can appreciate all those things as a scientist. And that forces me to confront more philosophical questions, such as what are we doing on this rock, hurtling through space? And do we have any right to believe that our presence here matters to any entity anywhere else? I don't believe it does. I think we are a magnificent creation, but I don't think we're the most magnificent creation on the planet—there are some other amazingly marvelous creatures that have evolved. We're pretty marvelous ourselves in some rather funny ways. But I don't believe that we're here for a reason. And I don't believe that it will matter whether we are here for three million years or for two billion years—I don't think it will matter one bit which one it turns out to be.

So how do we deal with our impact on the planet? One way to deal with it is to be totally selfish and say [that] during my lifetime, in a Western country with a reasonable standard of living, everything is definitely going to be fine. So why worry?

I am not able to do that. I think my mother taught me the difference between right and wrong a little too well. And so I can't take that very simple view, which lots and lots of people take. I've got a forest here. I could cut it down tomorrow, turn all those trees into dollars, put them in the bank, and I'll make more money than if I enjoy my forest and only harvest it sustainably. [But] I think it's morally wrong—seriously wrong—to cut that forest down just because I can make more money by putting the money in the bank. Because I think that forest has a right to be there. This is not science; it is purely philosophy. However, I think that it is philosophy that is more likely to enable us to learn how to tread lightly on the planet.

Toward the end of my academic career, I was talking to a neighbor one day as we were mowing our lawns. It was an early spring, and I made some comment to him about how we should expect this [kind of warming weather] in the future because of climate change. And he said, "Oh no, the majority of what I read says the climate scientists don't believe any of that stuff." And I said, "No, no, you've got it backwards." But he was convinced. He read the newspaper, he was an intelligent man, and as far as he could figure out there were a lot of scientists who were saying how there wasn't any evidence of climate change. And that was the stimulus to write the only nonacademic book I ever wrote.[1] I decided I would try to write a book that would set out the science in a way that if people wanted to, they would be able to read and understand it.

CHAPTER I

Coral Scientists between Hope and Despair

We are in a crisis that is too scary to confront and too important to ignore.
—Mary Pipher, *The Green Boat*[1]

THE PENDULUM EFFECT

I had just returned from the International Coral Reef Symposium in Hawaiʻi, where some 2,500 coral scientists, managers, and policy makers from seventy countries convened, as they do every four years, to discuss the present and future of tropical coral reefs. The conference was overwhelming. Waves of coral people from around the globe and dozens of simultaneous sessions made it a frenzied affair. One of the scientists I was trying to trace wrote in an e-mail, "P.S. yes I am at the symposium, running around like a headless chicken like everyone else." I took a moment to reflect on how unsettled I felt. More than the usual hustle and bustle of gigantic conferences that academics both despise and thrive on, the International Coral Reef Symposium was exceptionally exhausting. What was it about this particular conference that left me so discombobulated?

The answer came in a flash: what I experienced at the International Coral Reef Symposium was the bipolar oscillation between the extremes of hope and despair. Some of the sessions were dark and depressing and left me gasping for air, while others were either hopeful or altogether oblivious to the impending catastrophe that their co-sessions were predicting. It was like living in a split reality. This oscillation was made that much more severe in light of the "tsunami of urgent and life-threatening planetary changes"[2] that coral

---

FIGURE 7. Wide-grooved brain coral (*Colpophyllia natans*) near Jupiter, Florida, September 2012. Courtesy of Coral Morphologic.

scientists, like all of us, have been bombarded with—what scientist Will Steffen calls "The Great Acceleration."[3] But coral scientists are probably worse off than most of us in the face of such acceleration. Spending their entire lives studying organisms who are in the fast lane to massive death has been exposing them to the harshest dimensions of these challenging times.

This chapter is an exploration of the extremes of despair and hope. In many ways, it is a microcosm of the entire book, which will also swing between these extremes, but will strive to examine them more deeply. I will start the chapter by exploring the book's central argument regarding the relationship between hope and despair. I will then move to introduce coral restoration's hopes for the future, followed by an account of the depressing state of corals as evident from the recent monitoring of coral bleaching events. Next, I will discuss assisted evolution and experiments with "super corals," identifying what some argue is the genomic turn in coral conservation. In its final sections, the chapter will complicate the story of hope and despair, first by discussing the tensions between global and local in coral conservation and then by turning to a specific debate between two groups of conservation scientists, which I will refer to as the Cinner-Bruno debate. I will conclude with Ben Halpern's story, which exemplifies the central argument of this book that although hope and despair seem like diametric opposites that cannot exist simultaneously, and are indeed perceived as such by many of the scientists I interviewed, close-up explorations reveal their interconnectedness. I am particularly interested in these scientists' demonstrations of hope that are more rooted and less transient, and in what distinguishes them from mere wishing or passive expectation. These expressions of hope on behalf of many of the coral experts I interviewed are arguably deep enough to encompass despair and even draw on it as an inspiration for action.

"DARKNESS IS YOUR CANDLE": INITIATING HOPE

I will say up front that only after many transcriptions and observations, and upon reading and rereading the polarized narratives, did I begin to notice that each extreme contains the other, even as they are perceived by many coral scientists as mutually exclusive. As this chapter documents, certain coral scientists who were initially hopeful but then fell into the depths of despair seemed to eventually emerge with a more sustainable and empowered sense of hope than the one they had started out with.

This is confusing. Just as English lacks words for different kinds of snow, our vocabulary is inadequate for expressing the different types of hope evinced by life in the Anthropocene. Accordingly, several prominent scholars have chosen to abandon the term altogether in search of a more apt vocabulary.[4] But when browsing the current literature on the environment, I was able to find alternative interpretations of hope and of the relationship between hope and despair. One book in particular refers to the deeper type of hope as "active hope." Joanna Macy and Chris Johnstone write in that book: "The word *hope* has two different meanings. The first involves hopefulness, where our preferred outcome seems reasonably likely to happen. If we require this kind of hope before we commit ourselves to an action, our response gets blocked in areas where we don't rate our chances too high." Unlike this type of hope, Macy and Johnstone focus on *active hope*, which they define as "identifying the outcomes we hope for and then playing an active role in bringing them about."[5] Literary scholar Teresa Shewry asks along similar lines: "Could hope be a mode of facing rather than of mollifying or forgetting environmental loss?" Drawing on the work of anthropologist Anna Tsing, Shewry encourages intellectuals to move beyond their resignation to the current state of globalization and degradation toward recognizing the passions and dreams of justice that can fuel the struggles for more viable forms of environmental life.[6] While this type of hope is not naive, it also does not require optimism.

But I'm jumping ahead of myself. Before reconciling hope and despair, I would like to take a long moment—a book-long moment, in fact—to examine the ways in which these juxtaposed approaches underlie and even circumscribe the divergent projects in coral conservation. One of my earlier observations, which surfaced at the International Coral Reef Symposium, has been that many coral scientists do not experience hope and despair simultaneously, but rather choose one over the other for extended periods in their scientific careers or alternate between the two for shorter time spans.[7]

I also noted my own oscillation between the extremes. As I already mentioned in the Introduction, I spent the summers of my teenage years scuba diving in the Red Sea in Sinai, Egypt. But life took me away from the ocean, and so much happened to coral reefs in the meantime. I started to realize the extent of the change during my visit to a coral nursery in Culebra, Puerto Rico, in January 2015, when Edwin Hernández-Delgado showed me around his underwater coral nursery and spoke about his efforts in outplanting *Acropora* species from the nursery onto the devastated reefs (figure 8).

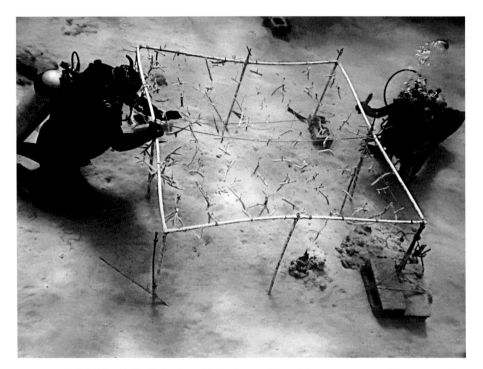

FIGURE 8. Edwin Hernández-Delgado and his colleague tie coral fragments to a metal structure using dental floss. Coral reef nursery, Culebra, Puerto Rico. Photo by author, January 12, 2015.

This firsthand realization that all is not well underwater inspired me to find out more about the state of corals. How have I not heard that they were in trouble, I remember asking myself, feeling like I had somehow betrayed their trust. Hernández-Delgado was able to put me in touch with Australian scientists when I visited the Great Barrier Reef in May 2015, and then with Israeli coral managers during my June 2015 visit to the Red Sea. Unfortunately, the sites at which I scuba dived in the northern Great Barrier Reef have since either bleached or died in what has been the most acute global bleaching event on record. Diving in the Red Sea was not very uplifting either. The sites I knew well have been closed to the public because of their degraded condition.

The process of writing this book has therefore not been easy. Overwhelmed by what I learned, I sometimes questioned the importance of the entire project, wondering if writing is an act of hope or a form of despair and whether my readers would even care. But the awed expression on my older daughter's face when she emerged from her first scuba dive in the Red Sea, as depleted as it already was, alongside the courage of so many of the scientists I have interviewed, have inspired me to go on. Writing emerged as an act of healing with the intention of making a difference—however small—for the sake of my two daughters, and for the awestruck child that still remained in me. In the tougher moments, I found solace in Sufi poetry from the thirteenth century: "Darkness is your candle. Your boundaries are your quest. . . . You must have shadow and light source both. Listen, and lay your head under the tree of awe."[8]

My struggles to cope with the massive death of corals reflect the situation for many coral scientists, who deal with this slow-motion catastrophe even more intensely, and at the same time must also make difficult decisions about what to do to protect them. When reading the barrage of depressing reports from the Great Barrier Reef, I wondered how, in the face of such devastation, these scientists find the motivation required to proceed with this hard work. "What's the point in saving corals if they will eventually die of global warming and ocean acidification anyway?" I remember asking Nilda Jiménez-Marrero, my first interviewee in this context and an employee of Puerto Rico's Department of Natural and Environmental Resources.

This was just before the third global bleaching event had started. Jiménez-Marrero's daily labor included tying corals together with dental floss to affix them to the ocean floor in an attempt to restore them from white-band disease, distinguishable by the white band of dead coral tissue that it forms. This disease has been identified as the central reason for the mass decline in the two dominant acroporid coral species in the Caribbean over the past forty years.[9] Jiménez-Marrero pointed out that the only alternative to dental floss was doing nothing—which is not something she could live with.[10] A couple of years later, marine biologist Nancy Knowlton of the Smithsonian Institution tweeted in a similar vein: "Bleaching is bad but giving up is worse."[11]

This kind of proactive approach made sense to me; it was even inspiring. Yet just a few interviews down the line, a scientist who described himself as "realistic"[12] swayed me to the other extreme. Instead of wasting time restoring coral fragments who will then die from similar stressors, he told me, we should focus our attention on the real change that needs to happen to reverse climate change: reducing global emissions.[13] And so I have been swinging, along with my interviewees, from hope to despair, from despair back to hope.

Dealing with such emotional volatility was hard enough. What I didn't expect, however, is the extent to which the two sides of the pendulum are disconnected from and even alien to one another. In fact, during a much-needed break at the Hawai'i symposium, an Australian ecologist complained to me about the three full sessions dedicated to the "unscientific" topic of restoration, criticizing what he perceived as its Pollyannaish undertones. Many other scientists I spoke with similarly went out of their way to explain why restoration is a mistake, claiming that it is a form of escapism that undermines coral conservation. By holding out hope that we may yet find a technological fix for climate change and other major environmental crises, they argued, humans (especially those living in developed countries) might come to believe that we don't need to introduce drastic changes to our ways of living in this world.

RESTORATION: HOPE

Sarah Frias-Torres grew up in Barcelona and spent her childhood on the Mediterranean. "I learned how to snorkel before I learned how to swim," she told me.[14] Frias-Torres transitioned from lab research in marine science to applied research in conservation and, finally, to restoration. She explained these transitions:

> When I began scuba diving as a marine biologist, the Mediterranean that I was seeing wasn't the one that Jacques Cousteau showed [on television]. It changed completely—there were no fish. And so that's what got me to say, "Okay, I need to work on conservation." But there comes a point when you realize that you've got to do more. So now I'm both in conservation and in restoration.

The central difference between conservation and restoration may not be readily apparent to an outsider: while conservationists have traditionally advocated for the preservation and protection of habitats and species with minimal human intervention, restoration ecologists advocate active engagement by humans.

Frias-Torres and I communicated extensively by e-mail. We finally ran into each other at the Hawai'i symposium's banquet dinner, which celebrated indigenous cultures. The singing, dancing, and clapping didn't stop Frias-Torres from outlining her detailed agenda for saving corals. "The restoration of the planet is the greatest challenge of our century, of our lifetime," she yelled over the noise. "If we don't work on it, there is no future." Frias-Torres's passionate belief in the power to restore was contagious, but she also seemed frustrated. Many of her colleagues from the traditional coral conservation world have been giving restoration folks a hard time, claiming that restoration is not a science, that it cannot be done at a scale that matters for corals, that it is taking both funding and attention away from real conservation, and that it is naive. She responded to each and every accusation (still yelling over the noise): restoration projects that do not adhere to scientific restoration principles have given restoration projects that do adhere to such principles a bad reputation; as a result of technological advances and given proper funding, restoration can increasingly be performed on an adequate scale; and restoration funding does not compete with conservation funding. Finally, Frias-Torres believes that it is conservationists who are being rigid and naive in that they are not willing to admit that the world has fundamentally changed and will never again return to its pristine state. She signed off with a rather controversial statement, the kind that might send shivers up the spines of many traditional conservation folks: "Nature is so far gone [that] you need to do restoration to jump-start it; it's not going to do it by itself."

The night prior to our exchange featured a roundtable with all the big names in coral conservation. During the Q&A period, Frias-Torres provoked the speakers and audience with a statement that I will quote in full:

> I learned this from a pediatrician working in poor villages in South America. He said [that] in the emergency room they don't have the resources to attend all the babies and little children at once. When in a triage situation, they only help the babies that are crying. Because if they are crying, it means they are strong enough [that] the medical treatment will be useful. For the babies who don't cry, there's no help. We are now in triage.... We can only save the coral reefs that are still crying. Crying for help. Because they still hold enough life that the conservation effort will be useful. What to do with the others? As a coral reef restoration scientist, I ask you: *Give us the rejects.* Give us the dead and the dying. We'll patch them up.... Conservation and restoration [should] work together. In this way, coral reefs still have a chance at survival.[15]

An uncomfortable silence followed. A few long seconds later, an unrelated question was asked and the conversation shifted away. The plea to accept restoration as a legitimate sister of conservation, rather than an unwanted child, was avoided—this time. The tensions surfaced again shortly after, when the facilitator asked the doom-and-gloom icon Ove Hoegh-Guldberg whether, in light of their precarious state, all strategies to save corals are now game. Hoegh-Guldberg responded with a graph: he argued that even if we balance out emissions, temperatures will continue to climb until mid-century at least. Only then, he continued, would it be time to embark on restoration projects. "But at the moment," he emphasized, "it makes no sense because you put [corals] into the ocean and they grow for a while and then bleaching comes along and they die." A member of the audience fiercely objected to this statement. The right moment to do restoration was yesterday, he insisted.

And so the debate between conservation and restoration continues, with no end in sight. In the meantime, however, between 2013 and 2015, Frias-Torres and her team cultivated forty thousand corals in thirteen ocean nurseries built in the African island nation of Seychelles in the Indian Ocean. After two freak events (a hurricane and an invasion by an encrusting sponge) killed about 7,500 of the nursery corals, Frias-Torres transplanted twenty-five thousand corals from thirty-four species onto half a hectare of degraded reef. She reported that the transplanted coral reef has significantly increased in fish numbers and that new baby corals have settled in the colony from elsewhere, a sure sign that "restoration is working."[16]

MONITORING BLEACHING: DESPAIR

After months of intensive aerial and underwater surveys (figure 9), in May 2016 researchers from the Australian Research Council released an initial estimate of the death toll from coral bleaching at the Great Barrier Reef. The massive bleaching killed 35 percent of corals on the northern and central sections of the 2,300-kilometer-long Great Barrier Reef system, the researchers reported in that month's issue of *Science*.[17] On twenty-four of the eighty-four reefs surveyed, 50 percent of the corals perished, including specimens that were fifty to a hundred years old. Such extensive coral bleaching on many of the world's reefs between 2014 and 2017 have highlighted the corals' vulnerability to thermal stress, with global warming currently "only" at 0.9°C.[18] After releasing this data, Jennifer Koss, NOAA's Coral Reef Conservation Program director, was quoted as saying: "We have boots on the ground and fins in the water to reduce local stressors. Local conservation buys us time, but it isn't enough. Globally, we need to better understand what actions we all can take to combat the effects of climate change."[19]

While predicted quite accurately, the extent of the third global bleaching event at the Great Barrier Reef and in other locations across the three ocean basins still caught many coral scientists by surprise. "It feels like a very special

FIGURE 9. Researchers survey bleached and dead corals in the shallow waters of Cygnet Bay in Kimberley, Western Australia, during the third global bleaching event. Credit: Christopher Cornwall, April 2016.

person whom you just started to get acquainted with suddenly dies on you," marine ecologist Joanie Kleypas told me about the coral colonies she has been working with in Costa Rica.[20] "I feel so helpless," she added. The third global bleaching event of 2014–2017 was caused by an El Niño event that coincided with higher-than-normal ocean temperatures due to climate change. The corals are extremely sensitive to higher water temperatures: exposure to just one or two degrees Celsius above the highest average summertime temperature for just four weeks (or for even shorter durations at higher temperatures) will result in heat stress that induces coral bleaching and is usually fatal. Until recently, mass bleaching events—those that affect corals in many areas around the globe—occurred about twice a decade, which provided sufficient time for coral recovery. But record-setting global temperatures over the past decade are causing more frequent episodes of coral bleaching.[21] Some scientists warn that if greenhouse gas emissions remain on their present trajectory, 90 percent of reefs may experience severe bleaching annually by 2055. At that rate, corals will likely not be able to recover.[22]

Coral Reef Watch was established by the U.S. government in 1989 to monitor coral ecosystems.[23] Using climate-monitoring satellite data, Coral Reef Watch informs marine park managers and scientists when corals might be at risk for bleaching. The general public can also subscribe to the "Satellite Bleaching Alert." Since subscribing to this alert in November 2015, I have been receiving weekly notifications about the state of bleaching in my sites of choice, which I randomly picked from NOAA's 227 locations around the globe.[24] I must admit that although I initially opened and read each e-mail, now I only glance quickly at the titles and move on. This reminds me of something that Nancy Knowlton—the "mother" of #OceanOptimism,[25] which has received more than fifty-nine million tweets since its initiation in 2014—told me over Skype:

> You present people with large problems and no solutions and they just go to the bar. Why get upset about something that you can't do anything about? There are lots of stories in scientific venues that I don't bother to read. I know it's bad news, with just more details about how bad it is. There's only so much doom and gloom that I can carry.[26]

Shortly after the International Coral Reef Symposium in Hawai'i, I attended a smaller workshop directed by Coral Reef Watch coordinator Mark Eakin and his Australian colleague William Skirving at the University of Hawai'i on the Big Island. The goal of this workshop was to familiarize scientists and community activists with the "products" that NOAA has made available for monitoring and predicting coral bleaching, and to obtain their input on these products. During the break, I approached Skirving. What follows is the gist of our conversation.

IB: What [in the monitoring project] are you proudest of?

WS: I guess the fact that we've described, or predicted, months in advance, this current bleaching event, where it would strike.

IB: In the Great Barrier Reef or everywhere?

WS: Everywhere. The outlook product [that] we have indicated that the northern Great Barrier Reef would bleach and I, for one, was saying I can't believe our own product.

IB: How long in advance did it predict this?

WS: Three or four months.

IB: And did it help that it predicted it? Did it change things?

WS: [What do] you mean?

IB: I mean in terms of management.

WS: Oh—no. This bleaching event has been documented better than any other bleaching event we've ever had. It mobilized all the efforts of documenting it. Will it change the way [reefs are] managed? [The Great Barrier Reef] is already very rigorously managed, so no, that won't change. When a bleaching event is underway they don't change the way they manage.

IB: But maybe [it will change] the Australian government's licensing of mining companies and all that?

WS: Ah, that aspect of things. Well, that is a different story altogether.

IB: Is it?

WS: Unfortunately, in Australia it's not much different than the United States. In effect, you've got the politicians working to get back into power. Their time horizon is short, so they're looking for short-term gains, not long-term gains. And the short-term gains are jobs, not climate change. And that's the big problem, the big question.[27]

Renowned for his work of documenting the extent of bleaching at the Great Barrier Reef, Terry Hughes said along the same lines:

> The main issue is, obviously, reducing greenhouse gas emissions. Here in Australia, that's very controversial, because our government is trying to prolong the export of coal. The Commonwealth Government of Australia has recently issued a lease for a new coal mine in Queensland. It will export its coal across the Great Barrier Reef, so shipping and dredging will all increase if this coal mine proceeds. Obviously, the last thing the Great Barrier Reef needs is more coal mines.[28]

As if to confirm Hughes's concerns, the Australian government lobbied against and censored UNESCO's 2016 climate change report, demanding that any mention of the Great Barrier Reef be left out.[29]

And while global warming is becoming a major, if not the major, part of the problem for reef-building corals around the world, many coral scientists have shied away from advocating for the relevant regulatory changes in fossil fuel emissions, which some of them perceive as a political topic that operates at a scale they have little influence on. Instead, these scientists prefer to focus on calculating, monitoring, and predicting. Notably, none of the coral reef scientists I spoke with had questioned the importance of monitoring, even if only for the sake of knowing and recording. They are, after all, scientists—not policy makers—some of them explained when I pointed this out.

Although it happened before the extent of the third global bleaching event was realized, my initial interview with Hoegh-Guldberg was probably the gloomiest of all, as befits his reputation.[30] "If we don't arrest ourselves, we're going to destroy ourselves, a bit like an alcoholic planet," he told me over Skype. "We're going to do all the worst things to ourselves and we'll have only ourselves to blame for it. But we won't quite die; we'll be a shadow of ourselves, of course."[31]

ASSISTED EVOLUTION: HOPE?

In the final leg of my trip to Hawai'i, I was invited by Ruth Gates, director of the Hawai'i Institute of Marine Biology and president of the International Society for Reef Studies, to spend the day with her on Coconut Island, off the shore of Oahu. Although she both studied and worked with Hoegh-Guldberg for many years, Gates's outlook is radically different. From her perspective, "we will either have a reef in the future or we won't. But if we continue down this trajectory and we do absolutely nothing to assist the system, it's likely that we'll have nothing."[32]

By "assist the system," Gates is referring to a radical proposal that she has developed with Madeleine van Oppen, senior principal research scientist at the Australian Institute of Marine Science, to captive-breed corals who have survived the bleaching events and who are more tolerant of climate change as a result. These are the "super corals" of the future, as the two women call them, and the actions of propagating and then introducing them into strategic locations in the reef are intended to accelerate natural processes that may currently be too slow for the rapid pace of human-induced climate change (figure 10).[33]

Gates is expressly collaborative and supportive of different attempts to save reefs, including restoration, and she doesn't shy away from confrontation over these issues. This is her account of a conversation she had with an internationally renowned coral scientist from the gloom-and-doom side of the pendulum:

> As I said to [him], "I hate to be blunt, but you think that the restoration and assisted evolution need your blessing, and you know what? The reality is that you are so behind the fast ball. Restoration is happening. Assisted evolution is happening. We are doing it. And guess what? If we don't, there are places that will

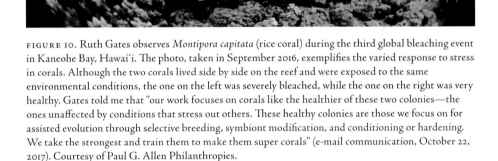

FIGURE 10. Ruth Gates observes *Montipora capitata* (rice coral) during the third global bleaching event in Kaneohe Bay, Hawai'i. The photo, taken in September 2016, exemplifies the varied response to stress in corals. Although the two corals lived side by side on the reef and were exposed to the same environmental conditions, the one on the left was severely bleached, while the one on the right was very healthy. Gates told me that "our work focuses on corals like the healthier of these two colonies—the ones unaffected by conditions that stress out others. These healthy colonies are those we focus on for assisted evolution through selective breeding, symbiont modification, and conditioning or hardening. We take the strongest and train them to make them super corals" (e-mail communication, October 22, 2017). Courtesy of Paul G. Allen Philanthropies.

have no reef left *now*. Christmas Island has no goddamn corals left. Are they going to wait for you to give them permission before they start doing these things? Absolutely not."[34]

Gates does have one major criticism toward the restoration community: she claims that they focus too little on coral genetics and epigenetics. "There's very little attention to the most robust genotypes," she told me, explaining that most nurseries focus instead on asexual reproduction. From Gates's perspective, the field of genetics represents hope for corals. In a follow-up interview from February 2016, she referred to this type of genetic work as "running our corals on environmental treadmills."[35] Her idea is to incrementally raise coral performance through selective breeding. Supported

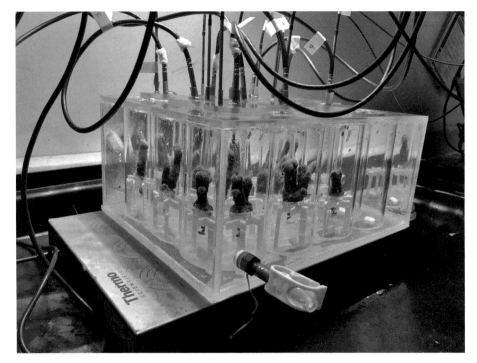

FIGURE 11. *Porites compressa* fragments are tested for respiration rates in a dark room at the Gates Coral Lab in the Hawai'i Institute of Marine Biology on Coconut Island, Hawai'i. Respiration is a central physiological parameter to measure coral resilience. Photo by Katie Barott, September 2016. Courtesy of Gates Coral Lab.

by a $4 million grant from Microsoft cofounder Paul Allen, Gates and van Oppen have been designing experiments to condition corals to stressful environmental conditions, with the aim of inducing epigenetically controlled stress tolerance that would then be passed on to the next generation[36] (figures 11 and 12).

Both Gates and van Oppen go out of their way to stress that their project is not about genetic modification.[37] Rather than tinkering with genes, they are merely doing what humans have been doing with wheat and tomatoes for centuries: crossbreeding for selected traits. "We have to be realistic about where we are," Gates told me. "For the most part, coral reefs are almost all declining. That means it's not sufficient to step back and wait. People tell me [that] nature might do that [by] itself. It might, but it doesn't seem to be. Not quickly enough."[38]

FIGURE 12. Dr. Nicole Webster (foreground, left) and graduate student Emmanuelle Botte (foreground, right) discuss the Evolution 21 experiment in the outdoor aquarium area of the National Sea Simulator (SeaSim) at the Australian Institute of Marine Science. Evolution 21 is a three- to five-year experiment held outdoors in natural lighting, exposing eight taxa (including the corals *Acropora loripes* and *Platygyra daedalea*) to three climate scenarios: current, mid-century, and end-of-century. Working in nine independent mesocosms (an experimental system replicating communities under controlled conditions), scientists control the temperature and the partial pressure of carbon dioxide. Photo by Christian Miller, May 2016. Courtesy of the Australian Institute of Marine Science.

A day with Gates meant an abundance of positive thinking. As we walked around the island, everyone we came across was delighted to see her. From the boat driver through the trash collectors to the institution's myriad volunteers, Gates greeted them all warmly and knew their life stories—and they responded with much appreciation. "People want to do something, they want to help. The gloom and doom is paralyzing, the scope of the climate change issue is so paralyzing." But not so for Gates. As befits her black belt in karate, she has readily taken on the challenges of climate change. And as with karate, she told me, most of the battle to save corals is mental, not physical. "I want to start a movement," she revealed in one of our many animated conversations. "There are people who do that for a living, who manage campaigns. I don't know how to do that. But that's what [saving corals] would entail. We need to empower people to do it."[39]

COMPLICATING HOPE AND DESPAIR: GLOBAL VERSUS LOCAL IN CORAL CONSERVATION

For the coral scientists I interviewed, not taking action isn't really an option. Instead, the relevant question for many of them is what actions they should be taking in the face of the deteriorating state of reef-building corals. The distinction between local and global actions has been central to this question. While some scientists, including Jeremy Jackson, argue that corals can and should be saved through local protections such as improving water quality and regulating fishing, others, including Ove Hoegh-Guldberg, insist that the local doesn't matter when global issues of climate change and ocean acidification are so overwhelming. For the latter, change is futile unless it addresses global emissions.

The debate between the local and the global underlies many of the discussions I have witnessed and recorded. For example, in a recent statement posted on the Coral-List, a NOAA-facilitated listserver networking nearly 10,000 coral enthusiasts from around the world,[40] coral scientist Dennis Hubbard admitted that "being a 'sophisticated scientist,' I often scoffed at local efforts." He continued: "I do understand that climate change 'trumps' all"; however, "I suggest that we stop trying to figure out which is the 'better' or 'not-so-good' strategy and applaud folks for trying."[41]

Seeking a nonscientist with whom I could share my observations about the hope-despair pendulum and its relationship to the debate about local versus global actions, I reached out to sociologist Elin Kelsey, who cofounded the #OceanOptimism project with Nancy Knowlton. Together, we contemplated the broader historical and psychological context of the schism between hope and despair in coral conservation. "The environment had become synonymous with doom and gloom," Kelsey explained.[42] Part of the problem, from her perspective, is the overly abstract approach often promoted by the dominant cadre of older, white, and male conservation scientists. This approach has led to a variety of emotional responses, she told me, including environmental fatigue, environmental despair and apathy, and environmental melancholia. Environmental catastrophe is a hegemonic discourse, Kelsey stressed. "I tried to write a children's book that was both hopeful and also based entirely on science," she recounted. "But it was very hard to get it published. The publishers literally told me that 'if it's a science book about the environment, it can't be hopeful.'"[43] The idea, she explained to me when we finally met in person one year later and were strolling through the English Gardens of Munich, is that if one is not depressed, one must be in denial.[44] Nancy Knowlton admitted that the depressing state of coral science has indeed caused many of her marine biology students to drop out. She attributed the decreasing enrollment in her classes to a growing pessimism about the future of corals.[45] During the course of this project, I have also heard about coral scientists who have experienced mental breakdowns and depression.

But one can resist this hegemonic environmental despair, Kelsey contended. "If you talk about things in more hopeful terms, you actually find that you're more focused on solutions and you're more likely to create opportunities to become more hopeful." Although she has often been accused of being Pollyannaish, she brushes such accusations off. "There's lots to be hopeful about. That doesn't mean there [aren't] lots of things to be worried about. It's an *and* situation, not an *or* situation."[46] This, precisely, is the way of thinking that #OceanOptimism is trying to promote by presenting a more feminine perspective that is focused on the particular and the local. "The whole doom-and-gloom thing is a huge leveler," Kelsey reflected.

> It's as if everything was great and [now] it's all wrecked. It's a very sweeping statement that does not reflect the incredible level of specificity and individuality that we now know exists not only in our ability to access the other-than-human world, but also in our understandings of that other world. There's a huge complexity of what motivates us and what makes us feel like we have self-efficacy, what gives us a sense of agency. . . . We want to celebrate the complexity of narratives, especially the ones that have an emotional understanding.[47]

Kelsey's situated and contextual approach recognizes the variations between coral conservation models applied in the Caribbean and those in Australia. Seen from this perspective, it makes perfect sense that I started my fieldwork at a coral nursery in the Caribbean. Because coral mortality in this region preceded that of the rest of the world by two decades at least, these were also the earliest sites of coral restoration projects. By contrast, coral conservation in Australia has mostly been managed by the Great Barrier Reef Marine Park Authority—a massive apparatus that has been widely applauded as the ultimate model for park management. Protection was the name of the game in Australia, and it seemed to be working—until the third global bleaching event hit the region in 2015–2017. During this relatively short period, some of the most protected and pristine areas of the park were also those most affected by bleaching and death. Certain Caribbean scientists I spoke to couldn't keep from making sarcastic remarks as they welcomed their Australian colleagues to the catastrophe club. "We have a hundred years over them with the historical legacy of worseness," one of the scientists told me. "But they're doing a great job. They're working hard."[48]

On a more sober note, another scientist explained the historical differences between the regions and why they have manifested in such divergent conservation models. "The Great Barrier Reef is huge," he said, reminding me that as the world's largest coral reef, this system comprises more than 2,900 individual reefs and some nine hundred islands stretching about 2,300 kilometers (1,400 miles). "Restoration will happen on a minute scale against the majesty of that huge system," he explained. The Caribbean, on the other hand, "is culturally divided and everyone thinks about the reef outside their hotel or their home." Clearly, despite the drastic changes that occur at the global level, geography and cultural differences are still important for understanding coral science and policy, which are far from being homogeneous.

THE CINNER-BRUNO DEBATE

The debate about the validity of local actions against the global backdrop of climate change has split the community of coral scientists for years. This debate—which also aligns with the purportedly juxtaposed attitudes of hope and despair—attracted considerable attention from the popular media[49] when July 2016 saw the publication of two ambitious studies in the prominent scientific magazine *Nature*.

The first study, by social scientist Joshua Cinner and thirty-nine colleagues,[50] documented fifteen "bright spots"—namely, sites where human management has led to successful conservation.[51] Cinner explained in our interview that "the predominant mindset among coral reef ecologists in terms of human interaction is a very Malthusian or neo-Malthusian one, where the number of people is the problem and the solution is to lock people away from the resource—and that's how we [will] make things better." But that becomes a self-fulfilling prophecy, he argued. "Every time you investigate things, the only thing you look for is human population density as any kind of indicator."[52] If you were to poll the scientists attending the International Coral Reef Symposium about what's a good reef, Cinner continued, they would say that "it's one that humans haven't affected yet and is pristine." The belief in pristine nature allows no room for humans, he argued, claiming that this is a very poor solution for the future. "So we mixed that up a little bit," Cinner told me. "We showed that a bright spot is somewhere that has more fish than it should, given the intense human impacts it was exposed to." In other words, Cinner and his colleagues created a model that expressed their expectations for what a fishery should be like at certain locations and then recorded both the positive and the negative deviations from that expectation. Cinner and his bright spots were the talk of the town during the International Coral Reef Symposium.

At the heels of the "bright spot" study, John Bruno and Abel Valdivia's article, "Coral Reef Degradation Is Not Correlated with Local Human Population Density,"[53] drew on quantitative surveys from 1,708 reefs around the world performed between 1996 and 2006. The article concluded that "local management alone cannot restore coral populations or increase the resilience of reefs to large-scale impacts," calling for "drastic and sustained cuts in carbon emissions."[54] Cinner adamantly criticized Bruno and Valdivia's approach. "Under the 'sacred nature' idea," he told me, "biodiversity conservation is the goal and the only thing that matters is the number of species." Within this worldview, "social ecological systems are a distraction because we need to stop the loss of species and people be damned, we need to just keep them away." This approach, Cinner clarified, is the opposite of his own, which sees humans as part of nature, rather than as alien and even detrimental to it.[55]

How are we to reconcile these two scientific studies, the one showing how humans can positively influence reef survival on the local level, the other demonstrating that local efforts are futile because corals will be doomed if we don't control our emissions? The bitter internal debate within the coral community was soon picked up by popular media. In an article published in the *Guardian* in August 2016, Johnny Langenheim tried to account for the differences: "The apparent contradictions here are partly due to each report analysing different types of data. . . . One was looking at fish, the other at coral cover."[56] In a further attempt at reconciliation, Bruno was recorded saying: "Although I doubt most coral bright spots are bright because of local management, I do agree that . . . we should throw all our conservation dollars to keep them that way as long as possible."[57] The Cinner-Bruno debate, with its extremes of hope and despair, thus also neatly aligns with global-local tensions. Finally, where coral scientists find themselves on the hope-despair continuum also correlates with their approach toward the human-nature divide. Being so foundational, their definition of nature will manifest in—and even lead to—different research directions and management preferences.

SPOTS OF HOPE

Early in my research for this book, I interviewed Margaret Miller, research ecologist at the Southeast Fisheries Science Center of NOAA. At that point, she was clearly not an enthusiastic supporter of restoration.[58] A year or so later, she approached me at the Hawai'i symposium. She said that she wanted to update me on her position, which had changed considerably since we last spoke. Could we communicate after the symposium? When I followed up a few days later, Miller sent me a long e-mail response. "The climate threats are so large and so overshadowing, sometimes I have to throw up my hands and give up," she wrote. Then she offered the following analogy:

> In corals, you have multiple threats occurring. Climate threats are one of them—one of the wheels on a car. The other [threats] are the other three wheels. If you eventually fixed climate change but didn't address any of the other threats, you would still have a car with flat wheels. I want to start working on the other threats we know how to fix right now—get those three tires pumped up and working perfectly so when we finally address climate change, we've got a working car. That's what keeps giving me hope: that I can do something right now, even though it seems so overwhelming. . . . That's how I sleep at night and don't feel completely depressed.[59]

Although she used it only as an analogy, I thought it somewhat ironic that a working car gives Miller hope about the future of corals.[60] Also notable about this comment is that it considers climate change as merely one, and not even the dominant, threat to corals in the Anthropocene. Yet most notable is the transformation of Miller's approach from despair to active hope, which, perhaps counterintuitively, has paralleled the declining state of reef corals.

Richard Vevers, founder and CEO of the Ocean Agency, would probably strongly disagree with Miller's opinion about the relative importance of climate change for coral conservation. Vevers is not a scientist; he started his career in marketing and advertising "just about anything: toilet rolls, car insurance."[61] Vevers shared with me how, one day, the triviality of marketing toilet paper suddenly dawned on him, and he decided to pair his media expertise and marketing talent with his love of scuba diving. The Ocean Agency, which he started in 2010 with other ex–advertising executives, not only created a first-of-its-kind global reef-mapping project,[62] but also partnered with Google Maps to present a "street-level" view of tropical reefs.[63] Speaking to me from a hotel room in New York City, between high-profile meetings with mayors and foundation directors, Vevers explained that most people don't care about corals, which they regard as rocks. The only hope for changing the public's perception, therefore, is to promote ocean literacy: to educate people about life underwater. His vision is that anyone should be able to see the ocean in their own living room, or to take an afternoon dive in Papua New Guinea from their smartphone. "Most people talk about climate warming, not *ocean* warming," he told me. "They have no idea that 93 percent of the heat of this planet is absorbed by the ocean and that the ocean controls the climate."[64]

But what he sees as the extreme ignorance of the public about how climate works is in fact what also makes Vevers hopeful, as "the ocean warming story is something that is relatively easy to communicate." The message is on the wall, he believes; it just takes bringing people to that wall for the shift in mindset to occur. "Although it's depressing that we're going to lose 90 percent of the reefs," he told me,

> we've got to turn that into [an] opportunity. And that opportunity is the big wake-up call that the underlying issue behind climate change isn't the atmosphere—where everyone's been looking—it's the ocean. The untold story of ocean warming is the most exciting part of all this because we can get action on climate change by completely changing the debate. I truly believe that in twenty years, we'll have broken the back of this transformation and people won't even really notice there was a problem in the first place.[65]

To get to that happy point, Vevers has recently embarked on something new: the 50 Reefs project. Collaborating with Ove Hoegh-Guldberg and others, he wants to shift the attention from bleaching toward life. "We know we can only save between 5 and 10 percent of the world's coral reefs," he explained. "So what we need to do is identify the ones we can save . . . [those] that are least vulnerable to ocean warming. Then, let's focus our protection efforts on the top fifty, and set those up as climate sanctuaries."[66] Unlike traditional projects of biodiversity conservation that focus on threatened corals, 50 Reefs will focus on those reef-building corals who are most likely to survive.[67]

LEARNING FROM BEN HALPERN'S JOURNEY

Our pain for the world not only alerts us to dangers but also reveals our profound caring. And this caring derives from our interconnectedness with all life. We need not fear it.

—Joanna Macy and Chris Johnstone, *Active Hope*[68]

Although approaching corals and their unfolding tragedy from very different disciplinary backgrounds and life trajectories, federal marine scientist Margaret Miller, social scientist Elin Kelsey, and media specialist Richard Vevers have each followed their own calling to achieve a balance between despair and hope. This balance, I would argue, moves beyond the oscillation between the extremes to a more persistent mode of hope that is rooted in action. I would like to end this chapter with one final story about hope and despair.

It took some time before Ben Halpern, a professor of marine ecology at the University of California, Santa Barbara, felt comfortable enough to tell me about the personal journey he has undergone as a coral scientist. "As I got into the field of marine conservation, I was passionate about wanting to save the world, like all young people are," he recounted. His transformation happened during the 2004 International Coral Reef Symposium in Japan. It was similar to the symposium in Hawai'i, he told me.

> You go to these talks and every talk is depressing and everything is doom and gloom. And I was sitting around at lunch with my colleagues and we were like, "This is ridiculous, we want to go see the last of the pristine places before they disappear." But as we had soon realized, we didn't know where the pristine places were. It's hard to map pristine places—it's much easier to map where we are having an impact. So we decided to map these. Where we were not having an impact— those would be the pristine places.[69]

This lunch epiphany launched what ended up being a ten-year project to build global maps and impact indexes for the oceans.[70]

The two profound findings of Halpern's project were, first, "that there is no place on the planet untouched by human activity, so there is no truly pristine place left," and, second, that more than 40 percent of the ocean is heavily impacted by human activity.[71] To measure impact levels, Halpern and his colleagues leveraged an unrelated study that examined seventeen ocean sites around the world, using historical records of fish catches as well as abundances of habitat and species. He explained:

> We went way back, compiling huge amounts of historical documents to show what these places used to look like. And so we were able to use that information to say: here is what, not a truly untouched, but a really close to pristine system looked like a few hundred years ago in these different places around the world. And then [here's] what it looks like now in a much more degraded state.

Referring to this comparative measure as Cumulative Human Impact,[72] Halpern's main objective was to figure out how to combine different kinds of measurements into a single index that calculates the overall impact of humans on specific habitats and locations. The first overall impact index was published in 2008. In 2015, Halpern redid the entire global assessment, this time focusing on degrees of improvement. "We were measuring where places are getting better or worse, as well as the human stressors or threats that are driving those changes," he told me. The team found that 13 percent of the ocean had slightly improved, but that a much larger percentage had deteriorated. The work procedure was highly involved. Because the calculations were conducted at a resolution of one square kilometer, and there are 350 million square kilometers of ocean, the scientists had to rely extensively on mathematical models. Still, they were able to ground truth the data from the seventeen locations for which they had historical records.

The revelation from this major undertaking was Halpern's finding that only 4 percent of today's oceans are not affected by humans. In light of this, Halpern decided to shift his focus: there were still 4 percent to be saved, he realized. Here is how he explained the flip in his perspective:

> A lot of people try to inspire change by giving a negative message, like "We're running out of time, we must do something." But then you get the pushback—if it's already doomed, what's the point of change? As a response, you start to flip to the other pole. "I need to find success stories, I need to find reasons for hope," you tell yourself. The patches [that indicate pristineness] are places that might provide some hope.

Halpern called these places "hope spots," to distinguish them from Josh Cinner's "bright spots." Hope spots have nothing to do with human management, he stressed. Quite the contrary, they indicate that "we haven't totally screwed the planet, there are still some [pristine] places left."

In retrospect, Halpern realized that underlying his hope spots project was the fundamental assumption that humans are bad. "That's the dichotomy," he reflected. "It's that humans can be either bad or good and so we have either hope or despair. These are the two poles of our mindset." Again, he shifted gears. "This isn't a fair representation of what's going on," he recognized. "To really understand what's going on here, you need to shift the focus into one that includes people as part of the system that you're tracking." Stemming from this realization, Halpern developed the Ocean Health Index, a uniform measure of the health of ocean ecosystems around the world.[73] "Can you use the ocean in a way that doesn't overstress it? If so, then you get a higher index score; if not, then you get a lower index score because you cannot treat the ocean sustainably." According to Halpern, the human impact index demonstrates that "if we do things well in terms of managing how we interact with the ocean, we can actually get things turned around in the right direction."

How does Halpern reconcile the Cumulative Human Impact index,[74] which is focused on identifying pristine areas with minimal human impact, with the Ocean Health Index[75]—a framework that encourages sustainable human use—without going off the deep end? Although he was initially amused by this question, Halpern quickly turned serious again. "It's tough," he admitted.

> You still hear a lot of stories about how things are not looking good. But then I know that humanity is incredibly adaptive and innovative and that we can figure things out if we put our minds to it. I know we can be there. But how we get there, that's the real challenge. I'm still searching for that balance.

Teresa Shewry suggests that hope is not willful blindness but a contingent response to a world characterized by injury and loss.[76] Halpern's concluding words are reflective of this realization: "For me, living and experiencing the fate of the oceans and our role in that fate has been a personal journey like I never would have expected. Where is the solution that will save the planet? What do we need to do?"

"This night will pass," Rumi instructed more than eight hundred years ago. "Then we have work to do."[77]

Prophet of Doom

An Interview with Ove Hoegh-Guldberg

Ove Hoegh-Guldberg is the inaugural director of the Global Change Institute and a professor of marine science at the University of Queensland, Australia. He has held academic positions at the University of California, Los Angeles, Stanford University, and the University of Sydney, and is a member of the Australian Climate Group and the Royal Society (London) Marine Advisory Network. In 1999, he was awarded the Eureka Prize for his scientific research. I interviewed Hoegh-Guldberg twice: once at the early stage of my fieldwork (February 25, 2015) and again more than two years later (May 22, 2017). I also met him in Waikiki, Hawai'i, on June 23, 2016. The following text is an edited compilation of our conversations. Hoegh-Guldberg has been cautioning about the impacts of climate change on coral reefs since the 1990s and has lobbied politicians on this front for many years. I couldn't envision writing a chapter on global bleaching without foregrounding his narrative.

IB: How did you become interested in coral reefs?

OHG: My mother and father are very much into nature. My mother used to take me down to a small river near where we lived in Sydney [Australia], equipped with a small fishing rod. I hated killing things, so I took the fish home and the fish became my pets. My Danish grandfather was a butterfly collector in his retirement, and he used to take me on expeditions, too. He'd come out from Denmark and we'd go to different places. One of the places he took me to, in 1969, was the Great Barrier Reef. I was ten years old when I saw corals for the first time. I spent this wonderful week and a half with my two elderly grandparents, snorkeling.

IB: Was he expecting to find butterflies in the water or—

OHG: Well, there were butterfly fish of course [laughs], but that wasn't the reason. He was collecting butterflies in the forest near the resort. And I was very much into collecting butterflies as well. Then I got fish tanks and collected little tiny tropical fish. So at the age of sixteen, I was spending all my time collecting fish. My parents almost moved out of their house because of the fish tanks. I was working at an oceanarium when I realized that you could actually go to the university and learn about these things. And, before I knew it, I was in my first university year at Sydney University as a research assistant for Peter Sale. We worked on an island in the Great Barrier Reef, and it was nirvana. I had arrived. It was beautiful. Every day I had to be hauled out of the water because the light was failing. And I've just been passionate ever since.

The following year I went to do my Ph.D. at UCLA. I worked with Len Muscatine, a famous coral biologist. Len had a massive influence on the field, and I was very lucky to be one of his last Ph.D. students. In the first year of my Ph.D., Len received samples of corals in the Caribbean that had started to undergo bleaching. Everyone was puzzled by it and by what it meant. No one knew anything. It was 1982–1983. Reefs in Florida went bone white, and people were sending samples to Len, asking what was going on. Was it a disease, was it natural? So I started to work on this particular issue, and by 1989 I had essentially done a Ph.D. on climate change without realizing it. I had conducted experiments to show that bleaching was related to temperature.

And that changed things for me. I was always a conservationist. And I loved diving. But it suddenly became clear to me that the thing I loved was now threatened. It didn't look good. In 1999, I published a paper that cautioned that reefs could disappear by 2050.[1] I did a very simple thing: I took the projections of future sea temperatures in the tropics, and then simply [compared them to] the temperatures in which corals would bleach and die. And when I put those two data sets together, I found that, basically, instead of having bleaching events every fifteen years, like we had since 1970, we would have them on an annual basis, and [that] they would be much more intense than what we had then, by the middle of the century. I remember being hammered by my colleagues, who were attacking me, behind my back and in front of me. "How could this be? This is ridiculous. Yes, we have a problem, but this is a long way off!" It became political.

Since then, I've evolved into a much different scientist. I spend most of my time liaising [with] politicians, working with NGOs, going global. I've started this institute at the university here that is looking into solutions for these big global questions. It's almost like suicide prevention.

IB: You call it suicide prevention—that's a bit catastrophic, isn't it?

OHG: Well it is, but there are solutions. I work very closely with the [Intergovernmental Panel on Climate Change] and have been leading their ocean efforts. I've become a different sort of person. I'm not so depressed because I'm involved with the solutions and I'm pushing those solutions and getting those things done. Some days I wake up, and it feels like the world is small and it's just a matter of changing some things. There is this sort of connectivity then that you start to see across the planet. There are solutions everywhere, it's just a matter of linking those things up and leading into a future that's really full of opportunity. Then there will be other days when I wake up and say, "Damn, it's so big." [The world has] seven billion people on it, and most people aren't aware of the climate catastrophe that we're heading toward, and so on and so forth. I mean, if we don't arrest ourselves, we're going to destroy ourselves. A bit like an alcoholic planet. We're going to do all the worst things to ourselves and we'll have only ourselves to blame for it. But we won't quite die, we'll be a shadow of ourselves of course.

IB: What amazes me is that a lot of conservationists say that it's all doomed, but then they act as if it's hopeful. I guess that's the only way to survive this period, mentally?

OHG: Yeah. And, unfortunately, I probably share the point of view that a lot of what we're doing in terms of conservation actions is futile until we stabilize the climate. I've been very involved in projects where we've grown coral back onto reefs in the Philippines using some very clever techniques. But, of course, if you haven't solved the problem, which is warmer seas and deteriorating water quality, you're putting coral communities there and they're there for about a year and then they die. So you need to solve [the larger] problems.

IB: Isn't this a kind of Sisyphean behavior?

OHG: Yes it is, and that's a good way of putting it. There's an interesting set of psychologies here. An important one is the psychology of the reef gardener who wants to keep gardening. [Many scientists] just can't see a future without coral reefs. They're denying the existence of a problem because they can't deal with it mentally.

IB: Now, why do you call them gardeners? Do you actually have to engage in intense management of these reefs even after you restore them?

OHG: Yes! If you put those beautiful plants down in your garden and then you go away for a couple of months, you come back and they'll be overrun by snails and all sorts of things. Because we've perturbed the system, it's not enough to just put the coral gardens down and walk away. You've got to then tend to them. Because you don't have the grazing fishes any more, you've got to pull up the seaweeds. So it's very much like a garden. In fact, it *is* gardening, it's underwater gardening.

There are places from which [corals] have already disappeared. It's not distributed evenly—some places have some, some places have none. And those places that have none, people are now trying to get in there and look at restoration techniques, and there's a whole industry that has been emerging around restoration. Unfortunately, I am the party pooper. [Like I said,] restoration won't work unless we stabilize the climate. I think we're wasting a lot of money doing this sort of thing. Not to say that we shouldn't be trying and refining the techniques and so on, but until we deal with the climate issue, [restoration] is futile. This is rearranging the chairs on the *Titanic* to get a better view. It's a sort of head down, bottom up, almost burying your head in the sand, you know? Let's just block out those horrible people, like me, who say it's all futile. "Lalalalalala, can't hear you!" I'm the preacher. The prophet of doom.

IB: So tell me, why are these changes in the ecosystems something we need to be concerned about? What's the problem with having seaweed instead of corals?

OHG: That's a big question—why are we worried? There are several levels at which that gets answered. At one level, there's the beauty of nature, the uniqueness of reefs, things that go back three hundred million years. [Do we want to] live in this sliver of time in which people are putting this

ecosystem to death? I often use the analogy of *The Starry Night* by van
Gogh. I say, well, what if we decided we're just going to burn this painting?
Get rid of it. There'd be a huge outcry because of its uniqueness. So why are
we doing that with species? You know, each of those is like an impressionist
artist's work: it has existed over time, it's never going to happen again, you
can't recreate it, [and] once it's gone, it's gone. So that's one reason to lament
[the major ecological crisis of our time]. The second reason to lament it is a
very practical one: reefs that degrade and become algal forests are not as
productive as those that are healthy, so the fish communities on degraded
reefs around the world are not as abundant as those on healthy coral reefs.
You lose half the species. There may be some species that benefit, but overall you get less protein per section of coastline. Coral reefs probably support
five hundred million people worldwide in terms of food and nutrition. And
when you degrade coral reefs, there's less to go around.

The third bit is that coral reefs represent a million species, and once you
lose coral reefs, you also lose those species, which are an enormous bio-
chemical heritage. In Australia, for example, there are shellfish that have a
very strong poison that can kill you. About ten different companies now
make pain and heart medicine from understanding the poison of this particular organism. And that's just one out of one million species. Another
area that I work in is genomics, and it's fascinating. We're getting to the
point now where we could really utilize this stuff very efficiently. But what
we're doing, instead, is trashing it.

IB: What would you say to those who would suggest preserving those
genotypes in non-wild situations, such as aquariums or gene banks?

OHG: Well of course, yes—and there's a very vibrant aquarium
industry. . . . Actually, it is part of the insurance population, if you like,
against the loss of corals elsewhere. . . . But it's interesting [to consider]
what would happen if we continue on our journey with climate change and
end up with the only corals left growing in aquariums.

IB: Even what we call today "the wild" will eventually become such a
managed site that it won't differ that much from that storage space, no?

OHG: Absolutely. This idea that we go out in nature to conserve things
back to the way they were is a gone concept; that changed with the industrial

revolution. What used to be weeds in our garden have now become wild-life. . . . We're beyond conservation as we used to know it. [Instead,] we're now in the game of trying to garden and manage this moving vista.

IB: When you described your fascination with coral reefs, you didn't speak so much about corals as important in themselves, except when you compared them to *Starry Night*, and then you did so indirectly. Could you say something about your relationship to the corals themselves?

OHG: [My] fascination with corals and their diversity developed when I was an undergraduate and wrote my honors project on symbiosis in corals. I was fascinated by the idea that this ancient symbiotic relationship lives in perfect harmony and that together, coral and algae are conquering the universe. There's a whole mutualistic vein to this that affected my psyche. I believe that there are great lessons there for every process we engage with. There is definitely a lesson about mutualism. I mean, mutualism is where we should be going—we should be trying to balance our relationships on this planet. When you do that, you get happiness.

IB: You also called it connectivity. I liked that term.

OHG: Yes, before we destroy our wonderful civilization, we are progressing toward a higher level of understanding and respect for balance. Really, two things have happened to us that destroyed the earth. One was Victorian England and the notion of survival of the fittest. . . . The second was post–World War II, when we had this unfettered rise of technology without consideration for balance. After those two shocks, we've moved away from a world in which we can expect to ever go back to balance. It will be a new world, a mixture of species that have not been seen on the planet ever before.

IB: So there are disruptions, but then a new balance is found, until another disruption happens and a new balance is sought. Is that what you mean?

OHG: If you watch a person stumbling and they're sort of falling forward [as they] manage to keep going—that's what we've got. We're no longer the sort of robust person that's standing perfectly balanced. The earth is stumbling on its journey, and what we have to do is make it stumble less. But how do we do that? The earth is on a treadmill, [and] the treadmill's going faster and faster.

IB: I see—thanks for the clarification. Could you perhaps say more about why your work focuses on corals? You were interested in symbiosis, but aren't there plenty of other examples of symbiotic relationships in the marine environment?

OHG: Well, corals are unique because they build geological structures that you can see from outer space. I was pretty captivated by this very simple organism that had lived for millions of years and had created limestone deposits which shaped things like the French and Italian Alps of today. All of these places are made of limestone built, essentially, from the symbiosis between corals and these tiny plants. There are other contributors to those formations, but a significant part of them were these amazing reef systems, which depended on symbiosis—you give me sunlight and organic energy, and I'll give you nutrients back. We need more people out there communicating about what is happening to corals and getting to everyone, from American dads to Ethiopian grandmothers to Swedish teens. That's the only way the world will change.

IB: Tropical reef-building corals have this symbiotic relationship, but not all corals do. So do you think nonsymbiotic corals should be part of this story?

OHG: Thousands of meters below the surface there are the *Lophelia* reefs, which are really important for fish habitat. The deep-sea corals are real specialists. At two thousand meters, it's dark, it's cold—it's probably six degrees Celsius—there's no light, you feed exclusively on particles, and you produce these very fragile skeletons that form a bit of their habitat, which is very extensive. But those corals are never going to replenish the warm-water reefs of the future, or mitigate the damage they have suffered. Just take the six degrees in temperature change: moving to warmer waters would kill these cold-water corals. The tropical corals are the ones that have one million species living on them. The *Lophelia* reefs of the deep, dark seas have very low diversity by comparison.

When you're in the warm tropics, 95 percent or more of the corals form a symbiosis with tiny algae, and those algae are really important. First we realized how the algae lived within the corals; the next revolution was when scientists started to realize that maybe there are bacteria living in harmony with the corals, too. That then led to our modern understanding of corals, what we call holobionts. You know, there's maybe as many as a hundred different organisms living in and around corals that determine whether corals are healthy or not. This understanding only emerged in the 1990s. Nobody was thinking about that before. Once we got to the point where we could take tiny amounts of DNA in and around corals, we actually found a large number of other things living in this association, which we roughly call a coral.

IB: I like the idea of seeing a coral as an association. The coral is like a network, an assemblage, right?

OHG: Right. It's not just a single animal, or even two animals, it's a continuum of organisms that are all living and surviving together. And that's going to be really crucial as we go forward in time, as we change things around this complex consortium. Because we're going to have to really understand what makes a healthy coral, and it appears that these bacteria may have a very important role in their health.

Recently, we've been working a lot on mesophotic coral communities—those corals that live right at the edge of the depths in which corals can actually survive with photosynthesis, [which is] at about one hundred meters. We've been exploring whether mesophotic corals could act as a refuge against climate change. Members of my team have been issuing high-definition population markers to figure out whether corals that live in the deep actually supply baby corals to the shallows. As it turns out, some do and some don't. After a climate catastrophe like the one that we're crafting, some corals in the deep may be the source of new corals in the shallows. But it looks like for a very large number of corals, if they tried to have their babies in the shallows those babies would die.

IB: You are referred to in the coral community as the "doom-and-gloom bloke."

OHG: Now vindicated.

IB: Can you tell me about this vindication?

OHG: Like I said, in 1999, I published research that suggested that all reefs could disappear by 2050. Everybody said that I was too gloomy. But as it turns out, I was right—unfortunately. And so we're now seeing bleaching events happening year after year, instead of once every fifteen years. This was completely predicted. Given the gravity of the situation, it's no wonderful thing that I was proven to be right, because it's the worst thing on the planet—it's a sure sign that we're losing the world's most biologically diverse ecosystem outside of rainforests. It's also a really bad sign that we are rapidly heading toward an ecological disaster, not only in coral reefs but also for the rest of the planet. Because of their conservative nature, scientists had underestimated the rate at which this is happening. It's actually happening faster than even I thought it would.

IB: So how do you suggest moving forward?

OHG: This is where the 50 Reefs project comes in, [which is] an idea . . . formulated during the [2015] Paris climate talks. First, we've got to stabilize ocean conditions as quickly as possible. That's number one. The second thing we've got to do is to recognize that even if we do achieve the Paris goals, we will lose 80 to 90 percent of the world's coral reefs, so we've got to plan for the future. Once we have stabilized ocean conditions, we need to have a healthy stock of corals still on the planet. There will be only 10 percent of what we had before, but these corals will reseed future reef systems. So it's a two-step process: stabilize ocean conditions . . . and identify those reefs that are likely to survive climate change and look after them. Nothing else makes sense.

And that's what the 50 Reefs project is about. This is an all-hands-on-deck moment: we need to get ahead and plan for the long term. There's nothing else you can do. If you don't solve the climate problem, no amount of genetic engineering will get you anywhere near having a coral reef again. Then do your best to identify those reefs that have the best chance of surviving to the future and make sure they aren't wiped out by local factors such as water quality and pollution sediments.

IB: How do you figure out, right now, what the important reefs of the future are going to be in what will probably be a very different ecosystem? How do you do that?

OHG: It's basically looking at as many properties as possible and optimizing [the process]. So you get reefs that are showing the slowest rate of change, and then you look to make sure you're preserving those reefs that are going to represent biological diversity, and then you look at the connectivity of those reefs, and at the end of the day you evolve an optimal set of reefs. Fifty is not the limit on this program. It could be that after the first fifty, we then move to the next fifty. And it's also not about saying that these reefs are the only ones you should preserve.

IB: So have you created a set of scientific criteria for choosing these fifty reefs?

OHG: We've taken twelve of the leading scientists on coral reefs and we're now looking at all these different types of questions. How do you isolate those fifty areas of the world that have a really good chance of surviving the coming climate change challenge? Because even with the Paris agreement, we still have to go up by half a degree in average global temperature. Just look at the wreck the first one degree has brought and you will realize that we're headed toward tough times. We need to make sure that we save as much coral as possible so that in 2050, when we're in our dirge, reefs will be regenerating again.

IB: So, at the end of the day, Mr. Doom and Gloom is actually optimistic?

OHG: I'm neither. I'm not crying, and I'm not shouting hallelujah. If we're going to turn this around, we need to get busy. On the one hand, I think there's a real good chance that this will work. On the other hand, there are so many surprises in this climate change business—[and] who knows what the next challenges [will] bring for corals . . . One thing's for sure: if we do nothing, we [will] lose coral reefs within twenty to forty years.

IB: You're not in favor of restoration before we stabilize the climate. But how is restoration different from choosing fifty reefs and protecting them right now?

OHG: It's so different. At the moment, if you go out and try to restore a reef, all those wonderful corals will get hammered by the next warming episode. So, restoration is not for today: it should be about doing the research we need to make restoration a possibility in the future. For our project, we want to choose those reefs that will survive. So we're not restoring reefs, we're trying to protect as much as we can for when the climate has stabilized. Those fifty reefs will still be exposed to climate impacts, but the key will be to make sure that those reefs that are showing the best signs of surviving are not being destroyed by non-climate factors, such as pollution, overfishing, and so on.

IB: How do you feel about genetically producing hardier corals, rather than identifying and protecting them as in your 50 Reefs project?

OHG: I think we need to try genetic engineering. . . . But we've got to be really careful not to pretend that we have a silver bullet, because there are no silver bullets. We will lose coral reefs in forty years if we don't act quickly. That is what I predicted all those years ago. Now we need to deploy a wide variety of solutions. Some of my colleagues may criticize people for acting, but I criticize them for not acting. We need to get out there and make mistakes so that we learn how to take an ecosystem through a climate shift. That's what we've got to do.

IB: You're not much of a doom-and-gloom guy. You're very disappointing [laughs].

OHG: [laughs] Oh, on the right day, you can certainly catch me gloomy. But then I think it through and realize that there will be an urgent shift to alternative technologies. What we're seeing with Trump in the American presidency is the last gasps of the fossil fuel industry. Yes, it will be terrible for about four years. [But] with the backdrop of increasing storm disturbances that will be harming Americans, and [with] a loss of coral reefs on a scale like you've never seen before, what will happen is that even the dumbest politician is going to get on board. And that's already going on. Let's take the Republicans. Five years ago, they weren't able to utter the words "climate change." But there's been a subtle shift in that party, and now most of the reasonable ones are going, "You know what, we can't deny it any longer. It's happening."

IB: It seems like the Australians have until recently been more focused on a protection scheme, whereas the Caribbean scientists have been working more on restoration and other technologies. Could you explain these regional differences in management approaches?

OHG: The Caribbean definitely felt the impacts of climate change first. It was hidden among ideas that it was about local pollution, but if you look back you can see that the disease outbreaks were probably triggered by warmer-than-normal temperatures. The Caribbean is a smaller sea—it has probably undergone more changes than the Pacific. The people in the Caribbean were living that change since the 1970s. Since then, they lost at least 80 percent of their coral populations. That came well before the big events that characterized many other parts of the world. So the Caribbean has been a bit of an experimental playground. And this has taught us a lot of salient lessons as we go forward. But the key point is that the Pacific is now showing impacts on that scale. We're probably about twenty years behind the Caribbean. These differences have shaped local perspectives. Up until 2016, we'd only had one bleaching event in Australia, and we lost maybe 5 or 10 percent total.

IB: In the past, Australian scientists veered toward more traditional conservation. Do you think they are now moving toward more interventionist efforts, like in the Caribbean?

OHG: Definitely. That's where we are headed in Australia—we're headed toward more intervention. It's the difference between trying to manage for the past, to restore what you had, versus developing an ecosystem that functions like a coral reef, but has different elements. Because things can't live where they used to live, they're going to be different. And so we may see the same old ecosystems in some places, but with different players. But we may also see novel ecosystems—things that have not existed before, like lionfish on Caribbean reef systems. And we need to manage for those. I know that in Australia people are now looking at management with a new lens. Rather than preserve the old-growth forest that just burned down and try to recreate it, we're moving toward maybe plantation corals, like plantation forests. These are the ways things are changing. We're not trying to restore the past. We're trying to create something that functions like a reef and provides ecosystem services like the reefs of the past, but is maybe composed of very different organisms than the ones that were there fifty years ago. Conservation is not about trying to get the past to come back. We need to design a strategy to preserve the diversity of coral reefs in these very unusual and dangerous times.

IB: Even if you don't know which diversity to focus on because it's going to change significantly?

OHG: That's right. So, one of the criteria would be a very broad portfolio of reef systems, so that you capture as many opportunities to get those reefs through and to the future. You see what I mean?

IB: Yes. Thanks. To wrap up, could you tell me how you feel about what has happened in the last thirty years, from a more emotional standpoint?

OHG: If you mean to ask if I am emotional about this, then yes, I am emotional. I was crying in the 1990s. But at a certain point, you get over that grief and you get on with finding solutions. This is the last bus on the line, our last chance to stabilize conditions in the ocean through switching away from fossil fuels within the next twenty years. There couldn't be a more important question in the history of our species than whether or not we destroy the planet's biosphere. And so this keeps me captivated and I probably will never retire because I wouldn't feel comfortable about retiring until we've done our utmost best to make sure that this nightmare doesn't end the way many scientists think it will. . . .

We had lots of time in the 1990s to put in place a transition to renewable technologies at a pace that wouldn't have been too painful. But we've left it right up to the last moment. So it's going to be a very painful transition, but it's still possible. Go another decade, and it won't be possible. And it won't be only corals that will be in trouble, it will be all of us.

IB: You were involved in the Paris Agreement. Can you tell me about it?

OHG: I decided that what we needed to do was to bring the issue of coral reefs to the attention of as many leaders as possible as they gathered in Paris. So we worked toward having exhibits [and] we had a big night at the museum of oceanography in Paris, where we brought together Sir David Attenborough, Sir Richard Branson, and various others. We talked about the fate of the Great Barrier Reef. Then, two months later, it was actually happening—the reef was dying. So I spent all that time trying to get the message across that this ecosystem, like many other ecosystems, in fact all other ecosystems, was doomed.

Science has to get outside of its ivory tower. And that's not to say that we don't do good science, but we must communicate it to everyone. As I said, it's that American dad, Ethiopian grandmother, and Swedish teen. Everybody on the planet has to understand this. So it's about getting outside our comfort zone and communicating our science effectively so that people understand the problems.

IB: When people understand, do they actually care?

OHG: They do. You get a lot of, dare I say, self-assured sort of businesspeople who think they don't need to listen to the science. But once you slow them down and get them to listen, they become really concerned. They become concerned that it's so desperate, and they become concerned they haven't heard about it before. . . . That's [an issue of] communication. Regular citizens need to take their households, their communities, and their electoral districts to zero carbon as quickly as possible. We need to reduce carbon dioxide and other greenhouse gases that we're putting into the atmosphere. That's the problem, and of course that's the solution as well.

CHAPTER 2

"And Then We Wept"

Coral Death on Record

One of the penalties of an ecological education is that one lives alone in a world of wounds. Much of the damage inflicted on land is quite invisible to laymen. An ecologist must either harden his shell and make believe that the consequences of science are none of his business, or he must be the doctor who sees the marks of death in a community that believes itself well and does not want to be told otherwise.

—Aldo Leopold, *Sand County Almanac*, 1949[1]

DEATH IN ACTION

On July 27, 2016, an urgent message circulated on the Coral-List, an electronic network of thousands of marine scientists and coral managers. The title read "[Coral-List] NEED ADVICE!!—ACTIVE MORTALITY EVENT AT FLOWER GARDENS!" Here is the e-mail in its entirety:

> I just got a call from Emma Hickerson, Research Coordinator for the Flower Garden Banks National Marine Sanctuary. She's on the East Flower Garden Bank and reports what appears to be an unprecedented mass dieoff of numerous species

FIGURE 13. The genus *Orbicella* consists of three sister species, all listed as threatened under the U.S. Endangered Species Act. Their similar morphologies initially led scientists to lump them into a single species, *Montastraea annularis*, which included three morphological types: bumpy, columnar, and massive. Published in 2012, the latest taxonomic revision established that the *Montastraea annularis* species complex forms a separate clade of three species within the genus *Orbicella*: *O. annularis* and *O. franksi* are commonly known as the boulder and lobed star corals, respectively, and *O. faveolata* is the mountainous star coral. This particular specimen was rescued from PortMiami, Florida, by Coral Morphologic, a Miami-based multimedia group, before the Army Corps of Engineers started dredging operations there. Courtesy of Coral Morphologic.

of corals, sea urchins, brittle stars, and sponges over a large area. She says large mats of tissue are sloughing off and there appear to be large bacterial mats on the bottom. Large amounts of material and haze in the water is making for virtually zero visibility in some places. Water temperature is 86F. She isn't certain of areal coverage, but believes it to be affecting the areas containing at least three mooring buoys. I believe these buoys are about 100 m from one another. . . . They will continue survey and do random drops to determine the extent.

1) Emma would like advice on collection and preservation of materials. They have ethanol, formalin, and freezer space on board, but I don't know about other details (jars, etc.)

2) Does anyone believe that an emergency closure is warranted to reduce spread of the problem by divers, or for any other reason?

3) They are disinfecting dive gear, I believe with a vinegar solution. Any additional suggestions on better or additional protocols?

Thanks very much for any advice. I can pass it along to the field crew via sat phone.

The e-mail was signed by Steve Gittings, science coordinator at the National Oceanic and Atmospheric Administration (NOAA) Office of National Marine Sanctuaries.

The U.S. Flower Garden Banks National Marine Sanctuary is located in the northwestern Gulf of Mexico, roughly 105 miles south of Port Arthur, Texas. The formation of offshore salt domes in the warm waters of the gulf and the relative isolation of this location provided a colonization site for corals, who started building reefs there roughly ten to fifteen thousand years ago. The impacted reef was considered one of the healthiest anywhere in the region.[2] But in summer 2016, divers and researchers found unprecedented numbers of dying corals and other invertebrates on large and separate patches of the reef. Officials reported extensive white mats covering corals and sponges (figure 14) and estimated the mortality of corals to be nearly 50 percent in some of the affected areas. Twelve miles away, the reefs of the West Flower Garden Bank remained vibrant, bathed in clear, blue water and free of the problem.[3]

Coral scientists from around the globe responded within seconds. "Les: Please quickly clarify how to preserve affected tissues for metagenomic analyses for Emma and her team," one scientist wrote. A few minutes later, Les Kaufman, whom I had interviewed just a few weeks earlier at the International Coral Reef Symposium in Hawai'i, responded from his lab at Boston University: "Oh, wow—the nightmare. It's like when something bad afoot

FIGURE 14. These whitened corals in the genus *Orbicella* were devastated by a large-scale mortality event at the East Flower Garden Bank, which is part of a U.S. marine sanctuary in the northwestern Gulf of Mexico. The area shown is five to six meters across. Credit: FGBNMS / G. P. Schmahl. Courtesy of NOAA.

finally reaches out to an actual family member." He quickly moved on to suggest a few courses of action:

> OK. Emma and GP—beside what you'd already mentioned, I suggest preserving affected tissues for metagenomic analysis (DNA and mRNA). . . . It doesn't sound like temperature, but of course thermal data will be important—if there are loggers it might be wise to download them just to make sure the data are secure. . . . I think a brief moratorium during close monitoring [is] justified as a cautionary move.

"This won't be fun," Kaufman signed off.

"A nightmare indeed; condolences from all of us who treasure the Flower Garden Bank reefs," another scientist responded. Theories about increased temperatures, decreased salinity, high oxygenation, poor water quality, possible oil spills or seeps, particular pathogens, or a combination of some or all of the above soon flooded the Coral-List. Dozens of suggestions about how to monitor such conditions or prevent them in the future, alongside countersuggestions, flew across the world at an amazing pace (enabled, perhaps ironically, by underwater cables crisscrossing the deep sea, which can be hazardous to cold-water corals—but that's for another book).[4] "At least a NOAA vessel should be tracking [the dead zone] movements around the Gulf this week," one scientist offered, implying that perhaps low oxygen concentration due to pollution and other factors was causing this damage. The skeptical response was quick to follow: "Unfortunately monitoring is often like that TV commercial featuring a uniformed man standing in the middle of a bank robbery. He is there to monitor and determine if indeed a robbery is taking place—not [to] stop it." The same scientist reflected on the proposal to expand the sanctuary status to nearby reefs to control fishing: "Other than preventing anchor damage and possibly controlling fishing, one has to wonder if sanctuary status could prevent what the Flower Garden Banks is presently experiencing." In July 2016, NOAA recommended that until the causes become clearer, "the public avoid diving, fishing, and boating activities" in the area in order to "prevent the transmission of whatever is causing the mass mortality to unaffected locations, but also [to] protect divers from ingesting what could be harmful pathogens or toxins."[5]

Still unraveling as I write, the massive mortality event at the Flower Garden Banks exemplifies the urgent challenges that coral scientists are facing in the rapidly transforming world of the ocean. It highlights the breadth of expertise that is required from these scientists in their everyday work with corals and, perhaps more importantly, the limits of this expertise and the need to contend with *not* knowing. This incident also brings to the surface debates within this scientific community about the role of monitoring and the efficacy of local legal tools, such as marine protection areas, when dealing with changes on the scale of global warming and ocean acidification. Highlighting the increasing importance of genetic work to coral conservation, the mortality incident finally lays bare the unfamiliarity of many coral scientists with coral genomics, something I noted a few months earlier during the conspicuously small sessions I attended on genetic topics at the Hawai'i symposium.

As an outsider who observed the Flower Garden Banks event unfolding, I also became keenly aware of how united this community of thousands of coral scientists from across the globe can be when it comes to catastrophe management. Despite the diversity of these scientists' expertise, professional backgrounds, and personal convictions, each one of the e-mail communications about the Flower Garden Banks conveyed an emotional response beyond the scientific one. And while they relayed much sadness, urgency, and confusion, not one of the messages expressed helplessness or suggested giving up on this site, or on corals more generally. More than anything else, the Flower Garden Banks provided a real-time example of "death in action," demonstrating how the dying out of corals rallies those scientists who have dedicated their lives to studying corals into a mode of accelerated action.

Typically, however, their actions stop there. As scientists, many of them explained to me, there is not much else they can do except to serve as the spokespersons for their corals: to give them voice, image, and numbers when their decline and death would otherwise be invisible to, and ignored by, policy makers and the general public. It is perhaps unsurprising, then, that many scientists on the brink have become preoccupied with monitoring and calculating the devastation of their beloved corals. Are these obsessive practices of monitoring an act of avoidance, a way for scientists to cope with their own impotence to affect meaningful political outcomes? Or is monitoring a powerful act in itself—a way of making knowledge and thereby rendering certain humans accountable, which in turn recruits others to the task of bringing about political change? In other words, is monitoring an act of despair, or one of hope? Either way, scientists emerge from these emergencies as the sole witnesses to coral death: monitoring, documenting, recording, measuring—and weeping. This chapter will explore the tragic aspects of these crises and, sometimes, their coralated inspirations.

CORAL BLEACHING

As it happened, I chose a very depressing time to study coral scientists: they haven't yet figured out how to put out one fire, when another calls them to the rescue. In this case, the bad news from the Gulf of Mexico was soon followed by calls for action from other emergencies around the planet. In light of this cascade of catastrophe, it is perhaps not so difficult to understand why, not even a year after this event, Les Kaufman had a hard time remembering which site of mortality in the Gulf of Mexico I was referring to—he needed to move on to manage the next incident and the next crisis. Not only do coral die-offs happen at an accelerating pace, but massive communications and database infrastructures also enable the rapid sharing of information among coral scientists about these events, leading to an accelerated volume of coral catastrophe data, accompanied by a sensory overload for the scientists involved.

While the coral mortality incident in the Gulf of Mexico was marked by a white mat of unknown substance, coral death is more typically preceded by the bleaching of the coral herself. To better understand bleaching, we must return to basic coral biology. Scientists refer to the reef-building coral as a *holobiont*, a holistic entity composed of an animal "host," algal symbionts (*Symbiodinium*), and bacterial, as well as other, microbes.[6] The algae are the primary producers of reefs: they convert sunlight to biomass. The symbiosis between algae and coral is thus the foundation of the reef food chain, what scientists also refer to as the "trophic pyramid." Temperature increases can cause the coral holobiont to lose her pigmented algae symbionts and turn white, a process that was initially referred to as "paling" and is widely known today as "bleaching"[7] (figure 15). Bleached corals often cannot build their skeleton fast enough to stay ahead of erosion. As a result, they are likely to die.[8]

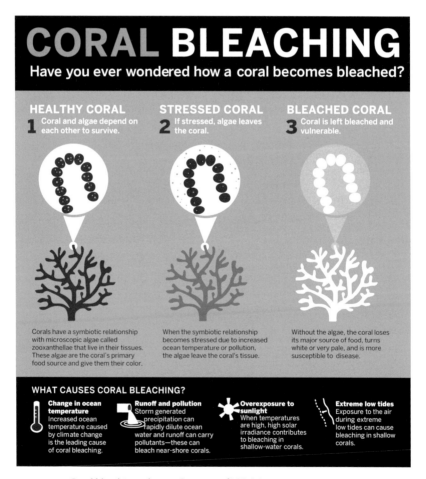

FIGURE 15. Coral bleaching schema. Courtesy of NOAA.

Bleaching is a relatively new phenomenon. It was first recorded in 1928, when marine scientists placed corals in a large opaque box to prove that they were not dependent on their symbionts. In 1964, Thomas F. Goreau published reports on the impacts of Hurricane Flora in Jamaica, which included several descriptions of bleaching. The first detailed analysis of this process was authored by Peter Glynn in the early 1980s. In 1985, Glynn further asserted that bleaching is likely linked to ocean warming.[9] Over the years, NOAA developed three categories for measuring bleaching severity based on geographic scale: local, mass, and global events. The first local bleaching events were recorded in the 1980s. A global bleaching event is defined by NOAA as a mass event (i.e., one that "involves bleaching across more than 100 square kilometers") that occurs through all three ocean basins—the Indian, the Pacific, and the Atlantic/Caribbean/Gulf of Mexico.[10] The first global bleaching event happened in 1998 and the second in 2010. In 2015, NOAA retroactively announced that the third major global bleaching event had begun in the latter half of 2014.

The third global bleaching event lasted three consecutive years—from June 2014 through May 2017—making it the longest and also the most severe bleaching event on record. As for the future, scientists now estimate that by the 2030s, more than 90 percent of the world's reefs will be threatened by local human activities, warming, and acidification, and nearly 60 percent will face high, very high, or critical threat levels.[11] "Catastrophic global warming has arrived," historian Iain McCalman exclaimed, even before the third global bleaching event reached the Great Barrier Reef, the focus of his study.[12] McCalman emphatically describes what bleaching entails:

> When corals are exposed to temperatures that are two or three degrees higher than their evolved maximum of eighty-eight degrees Fahrenheit, along with increased levels of sunlight, it's lethal. The powerhouse algae that live in the corals' tissues, providing their color and food through photosynthesis, begin to pump out oxygen at levels toxic to their polyp hosts. The corals must expel their symbiotic life supports or die. Row upon row of stark white skeletons are the result.[13]

Coral scientists are still debating the reasons for the expulsion of the symbiotic algae, many holding that the algae, not the animal "host," are the ones exercising agency in this process.[14] McCalman ends his book with a somber call for human agency: "It is a symbiosis which . . . has survived for some 240 million years, but which will split should those harsh forces so dictate. If anything can inspire us to prevent this, it's that very partnership itself, between two of the tiniest and most fragile creatures in the sea."[15] While the microscale of the algae-coral partnership is what had inspired McCalman, others may be more motivated to save corals for their role in enabling "higher" life forms such as fish and dolphins, or humans.

Despite their close relationship, coral bleaching and coral death are not synonymous (figure 16). Although a certain degree and duration of bleaching, accompanied by other accumulated threats, will likely result in coral death, the bleached coral may potentially recover.[16] The distinction between bleached and dead corals has been subject to considerable calculations and resulted, for example, in two separate categories in NOAA's coral monitoring system: "likely to bleach" and "likely to die." Mary Alice Coffroth, a marine biologist at the University at Buffalo, explains the difference:

> When the tissue is gone—that's death. [But] the end of the symbiotic element doesn't necessarily entail death, although that is likely to be the case if [the bleaching] is complete and prolonged. I sampled [corals in Florida] in May [2015] and they hadn't bleached; and then I went back down in the summer [of 2015] and they had bleached. . . . Losing the symbionts is a very bad thing. It's traumatic.[17]

The trauma of bleaching doesn't only affect corals, though. Coffroth told me that a friend and colleague of hers who has been researching brain corals in the Florida Keys has been deeply traumatized by the massive bleaching events that impacted the corals she has been studying and is currently dealing with mental depression as a result. Apparently, this response is not uncommon. According to psychologist Mary Pipher, "most of us are suffering from mild to severe mid-traumatic stress disorder."[18] Other writers have used the terms *planetary anguish*, *eco-angst*, and *ecological anxiety disorder* to describe the melancholia felt when one's environment is changing in ways that are profoundly negative.[19] "If you are paying some attention to the world, you are likely in some sort of pain," Pipher offers.[20]

CORAL IN THE COAL MINE

Scientists from around the globe agree that reef-building corals are dying en masse.[21] This highly visible death in the making—and the unique, two-stage, "double death"[22] of corals in particular—enables an extended time for pos-

Healthy - Dec 2014   Dying - Feb 2015   Dead - Aug 2015

FIGURE 16. The same reef in American Samoa before, during, and after the third global coral bleaching event. Courtesy of The Ocean Agency / XL Catlin Seaview Survey / Richard Vevers.

sible intervention when, as they bleach, corals teeter on the brink of their second death. The ghostly white spectacles of bleaching are the mesmerizing face of the encroaching mass death in tropical corals, their human-induced but self-orchestrated requiem (featured in figures 2, 9, 10, 16, 18, and 38). As such, corals have become an icon for ocean conservation. Their sensitivity to heat and the visibility of their bleaching make corals a model species for what our radically warming and polluted planet has in store for the rest of us, lending specific urgency to the scientific efforts to save them.

After the algae are expelled, some bleached corals will start to glow in bright fluorescent colors. "Like a supernova before a star's final collapse, the corals send out a steady stream of intense glow just before their inevitable demise," according to the *Smithsonian* magazine.[23] The reasons for this spectacular fluorescent glow in lime green, purple, pink, red, and blue are still being researched. Some scientists posit that the fluorescent pigments are the corals' sunscreen protection. More subtle when the algal symbionts are still around, the pigments intensify when the corals use them as a chemical sunscreen to try to protect their now naked flesh from the sun's harsh rays. The result is a sea of corals glowing bright in a rainbow of colors: "When you see it underwater, it's the most spectacular site in nature that you will ever come across."[24] After the corals' death, the glow stops; the bone-white corals are then gradually masked by a film of green and brown algae coating the ruins of the now deceased coral organisms.[25] "When you put the beautiful death next to the awful shots that show the aftermath of this death, you basically realize [that] coral reefs are screaming 'Notice Me!,'" Richard Vevers, director of the Ocean Agency, told me in our interview. "They are the canary in the coal mine," he added.[26]

I have heard the phrase "canary in the coal mine" numerous times, both from the scientists I interviewed and from media reports, and always felt slightly confused by it. Does the canary stand for the coral, whose death alerts humans that their own death is near? Or is the canary the bleaching events, which are signaling that the death of coral reefs is afoot? From Vevers's perspective, the canary represents the corals. For him, the corals are important insofar as they serve as a warning sign for the death of the ocean—and of us all. Unlike many of the scientists he has worked with, Vevers doesn't have a strong emotional bond with corals: he works with them because he recognizes their charisma for delivering a much broader message. In his words, "Coral death is our biggest weapon. . . . We need to use it to raise awareness of the climate issue. We need to milk everything we possibly can out of the ocean warming–related death of corals, especially [at] the Great Barrier Reef."

Prominent marine scientists and environmentalists from around the globe are similarly calling us to notice the widespread death of corals. "Coral reefs may be warning us to pay closer attention," wildlife biologist Doug Chadwick asserted in a 1999 *National Geographic* piece, which anticipated much of today's occurrences almost two decades ago.[27] Marine scientists see corals as measures of planetary health, proxies for predicting the global catastrophe of ocean warming and acidification and the even broader death of life as we know it on this planet.[28] Sylvia Earle, former chief scientist at NOAA, writes in her book *Sea Change*: "The living ocean drives planetary chemistry, governs climate and weather, and otherwise provides the cornerstone of the life-support system for all creatures on our planet. . . . If the sea is sick, we'll feel it. If it dies, we die. Our future and the state of the oceans are one."[29] As a recent news article put it, if the oceans *were not* absorbing more than 93 percent of excess heat captured by greenhouse gases, instead of today's average surface temperature of 59°F, the average global temperatures on land would be around 122°F—that is, unlivable. "More than 93 percent of climate change is out of sight and out of mind for most of us land-dwelling humans, but as the oceans continue to onboard all that heat, they're becoming unlivable themselves," the article warned.[30] Corals serve in these narratives as harbingers of doom, their very bodies performing the work of a doomsday clock.

Yet the differences between individual death, mass or global death, and even extinction are not straightforward. Elizabeth Kolbert, a staff writer for *The New Yorker*, attempts to capture these differences when she exclaims about the ongoing sixth mass extinction that "if extinction is a morbid topic, mass extinction is, well, massively so."[31] According to Jeremy Walker, a scholar of science and technology studies, mass extinction can be thought of as the end of evolution, an "anti-Genesis."[32] It is estimated that over 25 percent of the world's fish biodiversity, and between 9 and 12 percent of the world's total fisheries, are associated with coral reefs and would thus be adversely affected by their mass bleaching.[33] In other words, the mass extinction of many reef-building coral species will result in a greater mass extinction of all other reef-dependent species.[34]

CORAL REEF WATCH: SEEING DEATH

Many have pointed to conservation biology's extensive focus on catastrophe, and some have even referred to it as the "crisis discipline."[35] Coral scientists in particular are intensely focused on calculations of bleaching and rates of morbidity and on utilizing algorithmic models to predict future coral death and extinction.[36] This focus is also manifest in multiple managerial projects. One important type of project is the establishment of listing systems that focus on classifying species according to their predicted rate of extinction, what is referred to in legal and scientific language as "endangerment." I will discuss this type of project in chapter 4. A second project type, which I will discuss here, is the design of elaborate monitoring mechanisms to predict catastrophic coral events.

The Coral Reef Watch program was instituted in 1989 by the U.S. government and is one of the most comprehensive coral monitoring mechanisms in existence today. The mission of Coral Reef Watch is "to use remote sensing and onsite tools for near real-time and long-term monitoring, modeling, and reporting of physical environmental conditions of coral reef ecosystems."[37] The program uses satellite data to inform marine park managers and scientists when corals are at risk for bleaching. The reliance on climate-monitoring satellites is not incidental. "If fisheries management stood for the relevance of ocean science in the early twentieth century," wrote Stefan Helmreich, an anthropologist who studies the work of marine biologists in the deep sea, "climate monitoring plays that role now."[38]

With the advent of the Internet and other computational tools in the mid-1990s, Alan Strong, then director of Coral Reef Watch, developed satellite-derived sea surface temperature climatologies. NOAA defines climatologies as "charts that show the average conditions (or the 'climate') around the globe for each month of the year."[39] On the basis of data collected from coral reef scientists in the field, Coral Reef Watch calculated that corals begin to bleach when the sea surface temperature exceeds the average for the typically warmest month of the year by one degree Celsius for a certain duration, which I will discuss shortly. Using this data, Coral Reef Watch produces an online experimental chart that displays where such areas of high sea surface temperatures are found across the tropics in real time. This product, referred to as "HotSpot charts," is based on data obtained by satellites and made available online within a couple of hours. Today, Coral Reef Watch computers automatically generate HotSpot charts twice a week.

NOAA's coral monitoring system is based on "seeing" sea surface temperature. Mark Eakin, a biological oceanographer and the coordinator of Coral Reef Watch since 2005, told me in our interview that sea surface temperature "is one of the few direct measurements that we have." The satellite "sees the infrared radiation coming off the surface of the earth," and so "you're seeing the amount of infrared radiation or heat emanating out of the water."[40] Bleaching is predicted through observation of temperature changes at the top ocean layer.

The meaning of sight and vision in the ocean is worth a brief consideration here. In his book *Alien Ocean*, Helmreich points to the scientific connection of sight and light with knowledge: the word *theory*, which is associated with knowledge, is derived from the ancient Greek *theōrein*, "to look at" or "to contemplate." It is no surprise, Helmreich concludes, "that seeing through the opaque ocean has become the governing goal of oceanography, the grail of techniques of remote sensing."[41] Coral scientist Ben Halpern reflected, similarly, that "people connect to land because we can see it, we live in it, we can travel. The ocean is different: it's foreign, it's not where we live."[42] Although the ocean is invisible to many people, the visually stunning bleaching of corals has afforded us a clearer view of it than ever before. If polar bears provided the first conspicuous representations of the devastating effects of global warming on vulnerable species and ecosystems, now corals are the spectacular face of this crisis.

Legal scholar Patrick O'Malley argues that uncertainty "requires a certain kind of 'vision,'" which he refers to as "governing with foresight."[43] Such governance with foresight is precisely what Eakin was referring to when he told me that "using sea surface temperature gives us an ability to predict what's going on from one to three weeks in advance." This, he explained, is the result of the "time lag in the response of the corals to the temperatures."[44] Sea surface temperature is thus a good indicator of coral bleaching to come. The stress is calculated based on the average temperature in the warmest month of the year going back to 1985. "Corals adapt to the temperatures they normally see during the warm season," Eakin explained. Hence, temperatures above the maximum monthly mean are predicted to be stressful. The stress is also cumulative. This, according to Eakin, is how Coral Reef Watch calculates the accumulation:

> [If] you have one degree of stress that first week and the following week you have another degree of stress—you add those two together and it's two [degrees]. If in the third week temperatures rise, say, [by] two degrees above the maximum monthly mean, then in that week you get a two-degree week of stress and you add that to the previous two and now you're into four and that's how this accumulates.[45]

The next step in the prediction process is to translate the cumulative temperature stress into coral bleaching rates. Eakin clarified that "at four degree-weeks of stress you're likely to have significant bleaching; at eight, you're expected to have a widespread bleaching and significant mortality." Notably, this prediction doesn't take into account variations in coral species and ecosystems, resilience, or other distinguishing factors between particular reefs in specific geographies and times. Eakin explained that the system is currently not equipped to deal with such detail and is instead calibrated for rough predictions on a global scale.

In 2005, Coral Reef Watch added a new "product" to HotSpot, the "Satellite Bleaching Alert." Based on HotSpot levels, reef sites are designated with four levels of alert, ranging from lower to higher levels of certainty: Bleaching Watch, Bleaching Warning, Bleaching Alert Level 1, and Bleaching Alert Level 2.[46] An automated e-mail is sent to subscribers each time the alert

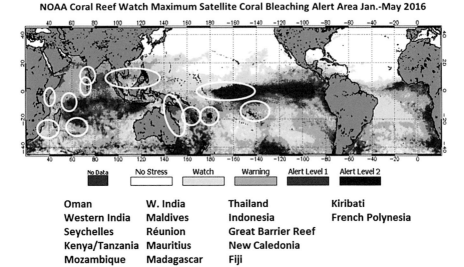

NOAA Coral Reef Watch Maximum Satellite Coral Bleaching Alert Area Jan.-May 2016

Oman	W. India	Thailand	Kiribati
Western India	Maldives	Indonesia	French Polynesia
Seychelles	Réunion	Great Barrier Reef	
Kenya/Tanzania	Mauritius	New Caledonia	
Mozambique	Madagascar	Fiji	

FIGURE 17. Screenshot of NOAA Coral Reef Watch map shows heat stress that impacted coral reefs around the world. The circles denote areas with reported severe bleaching in the first half of 2016 (5 km resolution). Courtesy of Mark Eakin, NOAA Coral Reef Watch.

status changes for any of the 227 sites around the globe. This service is free and available to the public.[47] Here is how Eakin explained the difference between some of these measurements:

> The HotSpot only tells you what's happening today, or actually yesterday. It doesn't tell you what's been accumulating over time. [For this purpose,] you have the Degree Heating Weeks. The [Satellite] Bleaching Alert is a simplified single graphic that combines information from both Degree Heating Week and the HotSpot charts. Rather than having a detailed scale, it breaks it down to alert areas and levels [to show] how critical the situation is for the corals.[48]

Launched in 2014, a geostationary satellite offers these spatial and temporal measurements more frequently and at an even higher resolution[49] (figure 17). The geostationary satellites can now produce sea surface temperature (SST) measurements as often as every fifteen minutes. Complementing this, each polar-orbiting satellite provides global coverage, including coverage for the regions missed by the geostationary satellites, by making near-polar orbits roughly fourteen times within a twenty-four-hour period. According to Coral Reef Watch scientists, "the combination of the six satellites provides the [five-kilometer] geo-polar blended night-only SST analysis with as many as fifty SST observations each night over the same location. These are then combined into a single SST analysis, for each pixel, each night."[50] Eakin described the benefits of this new technology: "The combination of those two types of satellites and the repeated observations every day allows us to have observations at a five-kilometer resolution, and each five-kilometer pixel has anywhere from ten to fifty times more data per day from the previous satellite [measures] at a larger resolution."[51] So, he concluded, "instead of getting one image, you're getting multiple images a day."[52]

The latest product developed by NOAA to monitor bleaching is a "satellite monitoring project in real time." Such real-time monitoring is perceived as more accurate than other kinds because it allows the user to see at the same time as the machine sees. Simultaneity is perceived to be a measure of truth, as highlighted by media reports of the dramatic bleaching events at the Great Barrier Reef.

Although the process of seeing sea surface temperature is portrayed as unmediated, multiple algorithms and computer processes are in fact deployed to translate this information into meaningful data and legible maps. The array of multimedia imaging and number-crunching algorithms is mind blowing, "a torrent of sense data that feels like a direct feed from what Kant once called the *mathematical sublime*, that domain of difficult-to-get-your-head-around measures and magnitudes."[53] The upper level of the ocean, its surface, is used to penetrate its depth in order to bring what is even more inaccessible into scientific view. Being the sole experts in reading these measurements, marine scientists thus assume the role of the oceans' exclusive spokespersons. Their ability to read the signs and monitor the state of the oceans provides the scientists with a sense of control that perhaps mitigates their helplessness in the face of the oceans' rapid changes. Communications professor and sociologist Chandra Mukerji's work on deep-sea science is relevant in this context. She observed that utilizing scientific techniques through the manipulation of equipment "gives scientists a way to assert their culture, and not become overwhelmed by the scale of the ocean."[54]

While typically operating at one global geographic scale, NOAA's calculations of time are based not on one but on *four* different future temporalities: near-real-time, the current season, past related climate patterns, and long-term climate models that predict decades and even centuries into the future.[55] The ultimate goal of the four-scaled temporal modalities is to forecast coral bleaching events—in other words, to render the catastrophe known and pre-dictable. In the words of science and technology studies (STS) scholars Claudia Aradau and Rens van Munster: "The objective, then, is to make the unknown known and show that what may seem unexpected in reality is an expectable outcome of causal processes."[56]

Novel documentation projects are mushrooming by the day as marine scientists place new instruments on planes, satellites, and even drones to gain a broader perspective on how well corals are doing.[57] As part of their ongoing attempt to "see" corals and their futures, scientists announced in 2016 that they will be outfitting a NASA airplane to map coral reefs based on the spectra of sunlight reflecting off reefs spread across the Pacific Ocean. This three-year, $15 million project, directed by Coral Reef Airborne Laboratory (CORAL), "will be the biggest and most detailed study yet of entire reefs, rather than just the small patches that scuba divers have been able to reach."[58] CORAL is part of a growing push to map reefs faster and in more detail than ever before. After surveying Hawai'i, the Great Barrier Reef, the Mariana Islands, and Palau, CORAL will have mapped about 4 percent of the world's reef area, hundreds of times more than any previous scuba surveys.[59]

Another project that has received recent media attention is the 100 Island Challenge. This project has already stitched together digital maps by using more than thirty-nine thousand photos of approximately forty-four thousand coral colonies near Hawai'i. In analyzing these digital reconstructions, scientists have been discovering distinct patterns of distribution.[60] The goal of the project is to inform and educate managers "about how their coral reefs work and what is needed to ensure that reefs persist into the future."[61]

THE GREAT BARRIER REEF: "AND THEN WE WEPT"

Terry Hughes is the director of the Australian Research Council Centre of Excellence for Coral Reef Studies at James Cook University in Townsville, Australia. Hughes is a protégé of Jeremy Jackson (whose interview follows this chapter) and is renowned for having trained and cultivated a strong cohort of next-generation coral scientists. He has been at the front lines of the extensive documentation of bleaching and, probably more than any other scientist, is perceived by both the media and the public as the spokesperson for the corals of the Great Barrier Reef. In the midst of the third global bleaching event, Hughes led a team to conduct aerial surveys of the Great Barrier Reef to assess the extent of bleaching in this region. He reported: "I've spent seven days in the air on a light plane and in a helicopter, crisscrossing the whole barrier reef. . . . When I've done my last flight, we'll have flown over about 900 individual reefs. We've scored every one for the severity of the bleaching."[62] Unlike in the case of NOAA's satellite project, sight and calculation occurred here through the unmediated, bird's-eye view of the scientist, who then proceeded to estimate the degree of bleaching in "real time."

Based on his naked-eye aerial assessment, alarming figures were produced that have contributed to a growing sense of crisis among the community of coral scientists. A 2017 article published in the news section of the journal *Science* reported that "mass bleaching has killed 35 percent of corals on the northern and central sections of the 2,300-kilometer-long system. On 24 of the 84 reefs surveyed, 50 percent of the corals have perished, including specimens that were 50 to 100 years old."[63] After completing the first 2016 aerial survey, Hughes tweeted: "I showed the results of aerial surveys of bleaching on the Great Barrier Reef to my students, and then we wept."[64]

The tweet's short ending encapsulates with almost biblical simplicity the immense pain that coral scientists are living through. "And then we wept" was how it felt when the director of the documentary film *Chasing Coral* showed scenes of coral bleaching at the Great Barrier Reef to a few dozen coral scientists during the Hawai'i symposium in 2016. I was the anthropologist in the room—and, loyal to this role, I started out in documentation mode, recorder on lap and camera in hand. But after a few minutes of being immersed in the collective sadness of coral scientists watching their beloved corals die, I couldn't help but weep with them, my recorder and camera all but forgotten.[65] The media was also taken by the grief of coral scientists. A *Washington Post* headline read, "'And Then We Wept': Scientists Say 93 Percent of the Great Barrier Reef Now Bleached."[66]

Although the bleached white skeletons of the dying-yet-not-dead corals (seen, for example, in figure 18) were visible from the air by plain sight, after the brownish algae smothered the dead corals, the reef's condition could be determined only by close-up inspection,[67] what scientists refer to as "surface monitoring" (Figure 9). Although most of the monitoring until that point had been conducted through aerial analysis, Hughes explained that "we [also] did

FIGURE 18. A turtle swims over bleached corals at Heron Island on the Great Barrier Reef, February 2016. Courtesy of The Ocean Agency / XL Catlin Seaview Survey / Richard Vevers.

an enormous amount of work underwater. In March [2016] we pretty much had one hundred people underwater on any one day. We went to 154 reefs on the Great Barrier Reef and we measured all sorts of things."[68] The main idea behind such measurements, according to Hughes, was to "ground truth" the aerial surveys, as he calls it. "We eyeballed two hundred thousand coral colonies and scored them individually into bleaching categories," he told a captivated audience at the symposium, explaining that "we've taken tissue samples to look at all sorts of metrics—cell death, genomics; we're looking at fish."[69] We came very close to the whole Great Barrier Reef being bright red, Hughes added. What saved the southern parts of the reef that year was the cyclone that cooled the water off. In his words: "We're now in a scenario where every El Niño event has the capacity to be a major bleaching event on the GBR unless we are lucky enough to have a cyclone. We used to worry about cyclones, now we hope for them. And if you think about it, that's a very precarious place to be."[70] The next generation of coral scientists may not even remember a time when cyclones were dreaded. Such generational amnesia is referred to in conservation science as the "shifting baseline syndrome."[71]

THE SHIFTING BASELINE SYNDROME
IN CORAL BLEACHING

In one of the sessions at the Hawai'i symposium, Terry Hughes emphasized the importance of recorded history at the Great Barrier Reef. In his words, "In 1998 and in 2002, the last severe bleaching events, Ray Berkelmans flew over most of the Great Barrier Reef marine park. He devised a scoring system for aerial surveys which is color coded. Green is good, yellow is not so good, orange and red are really bad. Those are the categories that he came up with."[72] In the latest round of global bleaching, Hughes and his colleagues drew upon Berkelmans's work. "We had an opportunity to redo that a third time," Hughes said, "and we've collected an enormous amount of data." Surveying more than 1,100 reefs over a period of several weeks, "we found three zones of bleaching. The top 800 kilometers or so above Port Douglas [were] 80 percent severely bleached" (figure 19).

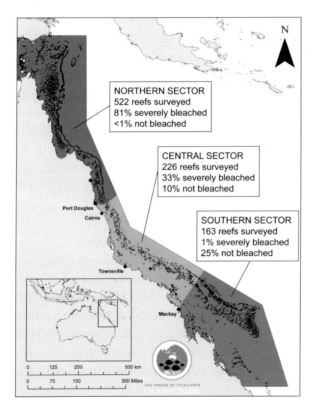

FIGURE 19. Map of the Great Barrier Reef, showing results of aerial surveys for a total of 911 reefs, April 2016. Courtesy of Australian Research Council Centre of Excellence for Coral Reef Studies.

Both aerial documentation from above and ground-truthing underwater have provided a comparative context through which scientists can conceive and comprehend the third global bleaching event and its futures. The colorful maps have been instructive in relaying some of the important characteristics of this bleaching event to the public and were reprinted by numerous media outlets. Nonetheless, Hughes explained that it would be difficult, if not misleading, to compare the 1998 and 2002 categories to those of 2016, and those of 2016 to others in the future. "There's been a shifting baseline of how we describe the severity of bleaching," he said at the symposium. "When Ray [Berkelmans] wrote his paper back in 1998, he categorized more than 10 percent bleaching as severe; now we think of 30–50 percent as not so bad."[73]

In other words, although the red color was used to represent the highest level of bleaching in both maps, this color does not indicate sameness in these two contexts. This comparison provides a good example of the *shifting baseline syndrome*, a term first introduced in 1995 by fisheries scientist Daniel Pauly to highlight the different perceptions of normality across generations.[74] The shifting baseline syndrome is particularly acute in the context of coral reefs because of the lack of reliable historical records about their status. Since there are so few established records to provide a comparative baseline, the level to which coral scientists are comparing the current state of corals is based on their personal knowledge, which is necessarily limited and shifting. Indeed, coral scientists of the older generation often point out that many of their younger students have never seen a healthy reef. Here is how Dennis Hubbard explained this phenomenon on the Coral-List:

> Younger researchers perceive "decline" on a different scale than those of us who were fortunate enough to see reefs in far better shape in general—and I'll bet our view of "pristine" is just as far off as theirs. Hindsight shows us that the decline had already begun in the "good old days" when we were just out of school. From a geological perspective, when I think of the "good old days," I'm thinking "early Holocene."[75]

The Ocean Agency—the nonprofit founded by Richard Vevers, mentioned earlier in the canary in the coal mine context—intends to tackle the shifting baseline syndrome, too. With expertise in media and photography, this team designed an underwater camera that would document reefs more accurately and efficiently than human divers. By August 2016, the Ocean Agency had already produced five hundred thousand images, all accessible to scientists through the Global Reef Record. "The camera has transformed the science," Vevers told me in our interview. He further explained:

> This provides unprecedented data for the scientific community. Producing a half million analyzed images that show different species in various regions around the world, all using the same protocol, offers the most valuable baseline of coral reefs, globally. For people to go back and revisit those baselines to understand what change has occurred is what we desperately need. This would be the first global study of coral reefs using a standard methodology.[76]

Such an accelerated production of photographic imagery again emphasizes the importance of vision for what is increasingly becoming a "spectacular" ocean science. "Obviously, the big issue [in the ocean] is that no one can see what is going on," Vevers reflected. This is how he came up with the idea of approaching Google to set up an underwater "street view" so that "everyone can go diving." In collaboration with the Ocean Agency, Google Street View includes underwater images from twenty-six countries. "It is our intention to survey all of the shallow reefs in the world," Vevers told me. He is hopeful that once humans don't have the privilege of shifting baselines and cross-generational amnesia, these new technologies will make people see, and thus change, their ways.

CORAL SCIENTISTS IN THE SPOTLIGHT

United in their grief for the Great Barrier Reef and for so many other sites of bleaching around the globe, coral scientists have nonetheless been arguing, behind the scenes, over how to present and interpret the bleaching information to the media and the public. The mainstream media has tended to highlight the gloomy predicaments. Here, from the *New York Times*: "Large Sections of the Great Barrier Reef Are Now Dead, Scientists Find."[77] And from the *Independent*: "More Than 90 Percent of Coral Reefs Will Die by 2050."[78] Although he was quoted extensively in these items, Hughes wasn't too happy with the headlines. In response to a flood of e-mail inquiries on the Coral-List, he clarified that he did not say that coral *reefs* were dead and dying, but that the coral *cover* was.

Created by living corals, the reefs themselves are calcium carbonate structures and quite dead already.[79] It is the live coral animal who loses her algae to the heating oceans, resulting in bleaching and then, possibly, in death. The media's proclamation of the reef's dying out was thus perceived by many coral scientists as factually misleading and left them perplexed. Why was their message not getting across? How can the public trust them if their message gets distorted? The scientists weren't sure what to do with this welcome but inaccurate coverage.

Another news item from that time announced the death of the Great Barrier Reef. The item, entitled "Obituary: The Great Barrier Reef (25 Million BC–2016),"[80] quickly went viral, spurring passionate discussions on the Coral-List. The scientists were particularly horrified by what they felt was a premature proclamation of death that, they feared, would result in the public's giving up on the Great Barrier Reef.[81] Les Kaufman from Boston University was also concerned, he told me, that the public would become very confused when scientists pronounce something dead and then, some time later, the same thing is declared dead again.[82] Underlying this discussion was another question: Is it morally justified to alarm the public in order to alert it to the dire state of corals, and of the oceans more generally? Some, like Ove Hoegh-Guldberg, have sounded this alarm as early as the 1990s. Yet many of his colleagues criticized him for doing so, calling him an "alarmist" and claiming that he sold his scientific soul to the media in exchange for headlines.

As noted earlier, Hoegh-Guldberg is probably the most outspoken "gloom-and-doom" coral scientist. He has persistently questioned the efficacy of pursuing local restoration efforts without first setting in motion a broader policy to address climate change. However, as the interchapter interview with him revealed, gloomy doesn't necessarily mean passive: "I'm not crying, and I'm not shouting hallelujah [either]. If we're going to turn this around, we need to get busy." According to Hoegh-Guldberg, conservation in the Anthropocene is about intention and action, not about optimism. Along the same lines, Macy and Johnstone write that active hope is a practice—"something we *do* rather than *have*" and that "the guiding impetus is intention; we choose what we aim to bring about [and] act for."[83] As he makes clear in the interview, these convictions have led Hoegh-Guldberg and several other scientists to shift from doing what they love, to doing what they feel they must do—that is, from research science to political action.

In the past, Hughes didn't always see eye-to-eye with Hoegh-Guldberg. But after the third global bleaching event, the two seemed to be finding themselves in agreement. At the Hawai'i symposium, Hughes told a standing-room-only audience:

I'm afraid I'm getting more and more skeptical about [local responses] as more and more places bleach for the first time. Reefs that never bleached or [those that] bleached lightly in previous years bleach just as badly as those that suffered in previous events. There is an urban myth out there that the corals are receding so they can recover in a decade and, don't worry, they'll come back. A 50-year-old coral takes at least that number of years to be replaced, so the northern part of the Great Barrier Reef is not going to look like it did in February [2016] again in any of our lifetimes.

"The scale of this [bleaching] event is about ten times bigger than a Category 5 cyclone," Hughes concluded. "It's like ten cyclones have held hands and walked between Cairns and New Guinea."[84] In another statement, he was even more explicit. While "local interventions might help foster recovery after bleaching," he said, "we can't climate-proof reefs by zoning them, by dealing with water quality, by restoring depleted fisheries. We've got to tackle emissions."

Such an exclusive focus on emissions has been characterized by some scholars as "climate determinism."[85] The problem highlighted by these scholars is that climate change is perceived as a unidimensional challenge that centers on how humans can achieve a reduction in their emissions of greenhouse gases in the coming few decades.[86] This is too simplistic and narrow, argues historian Dipesh Chakrabarty. Instead, he offers, we need to see climate change as one of a much broader set of complex problems in the Anthropocene.[87]

Despite his morbid predictions, Hughes remains hopeful. He was recorded saying to the *Huffington Post*: "The message should be that it's not too late, not that we should give it all up."[88] Famous for her ocean optimism efforts, Nancy Knowlton of the Smithsonian Institution assured the *Washington Post* that Hughes is not an alarmist.[89] Implied in her statement is that now even the more mainstream coral scientists are sounding the true alarm. The alarm is configured here as a moment in time that stands out; it is a wake-up call, a call for action.

MONITORING AND ACTION

What is the relationship between bleaching, monitoring, and meaningful action? My interviewees have characterized monitoring as either an act of hope or one of despair. Undertaking a hopeful narrative that emphasizes preparedness,[90] NOAA's website explains the practical importance of understanding bleaching: "Coral reef managers must be aware that a bleaching event is taking place, so they can act to protect corals from the long-term effects of bleaching. Herein lies the power of NOAA's Coral Reef Watch Program."[91] Indeed, in his role as Dr. Global Data and through his focus on monitoring temperatures and bleaching events on a global scale, Mark Eakin believes that the reduction of local stressors is key for saving corals. In his words, "You can help an organism survive one stress by reducing the other stresses at the same time to give them more of a fighting chance."[92] In the face of global catastrophe, such a focus on local action thus provides both meaning and motivation for preparatory monitoring projects.

In a less hopeful narrative, however, many scientists have expressed their frustration with the lack of response that has followed the bleaching predictions.[93] In this pessimistic interpretation of monitoring, the preoccupation of coral scientists with documenting bleaching is their way of keeping themselves busy with a false sense of action, which in fact is a form of tragic inaction—an impediment to, and a distraction from, political power. This recalls the analogy of the bank guard I quoted earlier in the context of the Flower Gardens: the guard is so preoccupied with monitoring that he misses the chance to prevent the robbery—as if he's there to determine whether a robbery is taking place, not to stop it. It may be worth noting, for the Foucauldians among us, that if conservation biopolitics is all about calculating and monitoring life in order to "make live," the connection between the two is severed when the project of monitoring life is divorced from the project of making life. This form of biopolitics not only questions the underlying project's active focus on life, but also possibly obscures the real project at hand: necropolitics, or the making of the living dead.[94]

So, what actually happens once scientists detect and announce that bleaching events are underway? In our interview, Eakin explained that "every country has its own laws and resource management regulations [and] will be making [its] own decisions. . . . A number of countries and local jurisdictions have bleaching management and response plans. Every U.S. coral reef jurisdiction, whether it's a state or territory, has a Bleaching Response Plan that gets triggered when an event [happens]."[95] The Great Barrier Reef Marine Park Authority's (GBRMPA) Bleaching Response Plan is an early and comprehensive example of such an action program. The GBRMPA plan includes procedures for prediction, ecological assessment, and communication of mass bleaching impacts, which consist of routine, responsive, and strategic tasks. According to *A Reef Manager's Guide to Coral Bleaching*, routine tasks "include the monitoring of environmental conditions and frequently updating assessments of bleaching risk." Responsive tasks include "rapid assessment of ecological impacts and increased communication activities," and "when bleaching thresholds are exceeded at multiple sites, a structured aerial survey is undertaken to determine the spatial extent and severity of bleaching in the region."[96] The main "plan for action" in the face of bleaching, it appears, is simply more monitoring.

In the meantime, nonprofit organizations have taken the lead in devising other, perhaps more active, plans for action. Ove Hoegh-Guldberg is the central scientist collaborator in a current program facilitated by the Ocean Agency to identify and protect fifty reefs on a global scale, a program I've already discussed.[97] Richard Vevers explained what has prompted his organization's shift from monitoring to action and the emergence of the 50 Reefs idea:

> For me, [2015] was a miserable year. We were talking about coral death and it felt like a huge yuck moment. I realized that this is a depressing job. It was at that stage that we came up with the idea of 50 Reefs. We accepted [that] we are going to lose 90 percent of corals and wrote that off as a fact, then [we] focused on the healthy areas and [the need to] protect them on a local level. This is a positive project where I feel we can make a difference.[98]

Hoegh-Guldberg clarified in our interview that 50 Reefs is not a restoration project, but plain old-fashioned preservation. Still, this project has been highly contentious among coral scientists. One prominent scientist wrote in an e-mail to the Coral-List in March 2017: "I find the '50 reef' concept morally repugnant—who among us has the right to tell hundreds of millions of people that their particular reefs aren't important enough to be among the tiny minority of reefs that will be 'saved' by dubious promises of restoration? We should try to save all reefs by dealing with climate change."[99]

Either way, many would suggest that calls for action like 50 Reefs are the exception rather than the rule. In 2016, Hawai'i saw heightened bleaching levels for the second summer in a row. The effort at that point, according to Eakin, was "to go out and make observations to see how severe it is, how it relates to the predictions, and look at the severity in different areas hoping to find some areas where the corals are protected by local currents or more resilient to the warming and impacts of bleaching events."[100] Here, again, the response was limited to observing and recording. Similarly, while the response plans I read through included a section entitled "Action," most of the actions listed were in fact restricted to monitoring and observation.

BREAKING THE RECORDS

The projections of global temperature to mid-century and beyond, even under the most optimistic scenarios, do not encourage any optimism. Somehow, we scientists have to report what is occurring, and continue to get up and go to the lab each morning. No other group of environmental scientists has had to face the *global* degradation, and likely disappearance, of the ecosystem they have been studying. And it all happened quite recently! Changing the trajectory is largely in the hands of others—governments, energy sectors, et cetera. This is the special, and disturbing, world that reef scientists find themselves in.

—Peter Sale, marine ecologist, 2017[101]

As befits a story about bleaching, death, and catastrophic predictions, this chapter will end on a rather gloomy note. Throughout the chapter, I have wondered about the role of monitoring and whether it is a form of action or a way to mask inaction. While accurate monitoring and documenting is a necessary foundation for any meaningful political action, one must also consider why such extensive monitoring projects have, for the most part, not translated into political action. Quite the contrary: the real political story seems to be how the expert knowledge of scientists has been dismissed as both illegible and alarmist (as contradictory as that may seem) and has thereby been neutralized and prevented from having political impact, therefore not leading to the remedial action called for. While a growing awareness of global coral bleaching has engendered elaborate monitoring projects that allow scientists to anticipate the imminent catastrophes and to predict the forthcoming ones, it has not led to the reduction of fossil fuel extraction, combustion, or licenses to pollute.

Coral scientists are united in their grief and in their commitment to save corals. Nonetheless, this chapter has also aired the heated debates among them. Alongside their disagreements on how to act in the face of the accelerated changes, which were predicted but also took them by surprise, scientists have been debating their relationship to the media and the lay public. What kind of an alarm can they sound without being tagged as alarmists? How dead should a coral reef be before it is pronounced dead? Those are just a couple of the questions that arose in the aftermath of the third global bleaching event.

In June 2017, the article "Coral Reefs in the Anthropocene" was published in the journal *Nature*.[102] Terry Hughes, who has been at the forefront of the detailed documentation of bleaching in the Great Barrier Reef and wept with his students in the face of these changes, is the first author of the article, which does not shy away from recognizing that "it is no longer possible to restore coral reefs to their past configurations."[103] The authors—including Josh Cinner (recall his "bright spots" work from chapter 1) and Steve Palumbi (who studies resilient corals; see chapter 3)—argue that local factors such as management and protection did not help the corals who were exposed to extreme temperatures and that an immediate change in climate policy is therefore required.[104] Emerging from these depressing realizations, the article ends with an emotional call: "We should not give up hope for the persistence of Earth's coral reefs."[105]

The Pristine Is Gone

An Interview with Jeremy Jackson

Jeremy Bradford Cook Jackson is an American marine ecologist and paleontologist, a professor emeritus at the Scripps Institution of Oceanography, and a senior scientist emeritus at the Smithsonian Tropical Research Institute in the Republic of Panama. He has published over 150 scientific articles—including eighteen in the prestigious journal *Science*—and has written or edited seven books. His articles are among the most cited in marine ecology. He has written and lectured about the "brave new ocean"—where clear and productive coastal seas turn into oxygen-starved dead zones and thriving food webs degrade to seas of slime and disease. He is also a proponent of the "shifting baselines" concept and was featured in the films *Before the Flood* and *The 11th Hour*. The text below rearranges and edits two separate interviews that I held with Jackson via Skype on February 20 and March 16, 2017. The interviews were preceded by an informal conversation at the International Coral Reef Symposium.

IB: Tell me a little about your professional trajectory. How did you get to be who and where you are?

JJ: I never started out to be a scientist. I thought I wanted to be a historian like my father. My Ph.D. is in geology, I did my course work in paleontology, and I sort of invented marine historical ecology. So, the historian ethos is there, except through the eyes of a biologist and geologist. I got thrown out of school, I went and got a job, I worked my way through George

Washington University as a night student, and then I fell in love with oceanography and paleontology. I had never been interested in those things, but I took a couple of courses that, as they say, changed my life. I wanted to be a paleontologist, but in order to do that I needed to be a geologist, so I stayed at George Washington University [and] got a master's degree in geology. I got my Ph.D. in geology from Yale. The great ecologist G. Evelyn Hutchinson was on my committee.

IB: Can you tell me about your Ph.D. research?

JJ: I wanted to [study the] present as a key to understanding the past. So I wrote a Ph.D. thesis looking at the mollusks that live in seagrass communities because that's a prominent kind of ecosystem that goes back over one hundred million years in the fossil record. I traveled around the Caribbean digging up and very rigorously studying the composition of the communities of mollusks that lived in these environments. I traveled with this wacky, wonderful guy who used to play rhythm guitar with the Jefferson Airplane and was doing his Ph.D. on the ecology of gorgonians and we had a ball prowling around the region. And we went to Puerto Rico and we went to Curaçao and we went to Panama and then we went to Jamaica where he was already doing his thesis and I met this amazing man, Tom Goreau [F.], who had founded this thing called the Discovery Bay Marine Laboratory, which was a shack on the beach, and it was the epicenter of global coral reef studies. It was *the* place in the whole world to be studying coral reefs. And those of us who got our Ph.D.s then [became] sort of a who's-who of coral reef ecology. . . .

A few years later, after I finished my Ph.D., the Australians were coming to Discovery Bay to learn what we were doing so they could take it back to Australia. And now they're the global leaders. At that time, Jamaica was beautiful. It was overfished to hell, but the reefs were the most magnificently beautiful coral reefs in the world. I was studying these clams that lived in mud and seagrass. But all my buddies were studying the sponges and the gorgonians and the corals and I would help them and they would help me. We were a community and we were living in this remote part of northern Jamaica. It was the most magical period in my life.[1]

I would dive with all my friends who were working on corals and stuff. A lot of what I was doing was just looking. Sort of like the old naturalists, you know, like Wallace in Malaysia or Darwin, where you don't have a lot of time and you see things and you need to exercise the ability to detect pattern and, from that pattern, infer process. The ultimate genius example of that is Darwin's theory of atolls and coral reefs. . . . That's genius, right? That's the ability to see patterns. I'm not a Darwin, but I'm pretty good.

IB: So what exactly did you see when you dove in the reefs?

JJ: Swimming around in these reefs, I began to notice that when a coral was growing next to a sponge, around the sponge there would be a bare area, like the DMZ between North and South Korea. And it was everywhere! Why was it there? . . . I never worked on corals, but I discovered these communities of organisms that lived under corals. And they were nature, red in tooth and claw; they were competing with each other and 100 percent of the space was occupied. I had seen all this stuff going on and I thought, "God, there's chemical warfare going on in here." There was this kind of intransitivity: A would beat B, B would beat C, C would beat D, but D would beat A. Which meant that, instead of having some dominance hierarchy with a King of the Mountain so that everything would be only one species, this intransitivity created a situation for the maintenance of diversity. I wrote a very important paper— say I with no shame—with an undergraduate student, Leo Buss, which was called "Allelopathy and Spatial Competition among Coral Reef Invertebrates."[2] And when we published that paper in 1975, Bob Paine, a very famous ecologist who was the guru of dominance, reviewed [it]. And instead of saying, "Squish him," Bob wrote an article saying, "You know, we have this worldview and it really works, but there's this really cool paper that we have to think about."

In simple systems, there tend to be clear dominance hierarchies. And the only way you have diversity is from what ecologists call disturbance, which can be a predator or logs banging into the shore and knocking the mussels. It's fundamental, it's absolutely basic to understanding what maintains the extraordinary diversity in the tropics. But this paper from 1975 was the first data expression that competition was nontransitive in the wild: there wasn't a perfect hierarchy. It was an empirical inkling that the ocean world was not hierarchical, it wasn't macho. So, here I am, I don't study corals, but I've become a world-famous coral reef scientist.

IB: Looking back, what was the event that most shaped your career as a coral reef scientist?

JJ: My transformative moment was after Hurricane Allen, when I suddenly realized what was happening. We were there, in Discovery Bay, in the summer of 1980 when Hurricane Allen happened. We had thirty-five people at our house. And you could smell the death the next morning. Islands created from the beautiful forests of elkhorn coral were turned into rubble islands. The stench of dead coral was everywhere. We had gotten everything out of the lab, [went] in the water the next day, and just started looking. These were the best-studied coral reefs in the world. So we had all this "before" data, and then this horrible hurricane came. It was the third-strongest hurricane in history at the time. It had two-hundred-mile-per-hour winds, the waves were forty feet high, [and] they destroyed the reef down to eighty or ninety feet. It was incredible. And there we were, the best and the brightest, studying the thing. I called up the National Science Foundation and said, "I want to divert twenty-five thousand dollars of my funds to bring people to follow this up." They said great, and even gave me a little extra money. We did it. And we published the paper, which I helped write with Nancy Knowlton.[3] And it is still a citation classic. Based on all this work we'd been doing, we predicted in the paper how the reefs would recover.

But they didn't. And that was the epiphany. Because why didn't they recover? Well, a couple of years later the sea urchin *Diadema* all died, the algae started to take over the reef, [and then] coral disease started to hit. We know from the fossil record that hurricanes happened all the time and the reefs would recover and you can see a succession [in fossil reefs] just as you see in a clearing when a tree falls in a forest. And so I thought about it and I thought about it and [I realized that] there was this oil spill, which affected a reef that had been studied by the Smithsonian in Panama for twenty years. I remember the Texaco lawyers saying, "You can't prove anything."

Because it wasn't a planned study, we had to use space as a substitute for time: I went and found nearby places that were similar to the places that had been affected by the oil spill. And those became my "controls," so I could compare [the oiled reefs versus the unoiled reefs] over time. [We] published a paper in *Science*, and I'm told that it led the National Academy of Sciences to recommend a drilling ban on the west coast of Florida.[4] This paper is still highly relevant. The next year we went out and all the corals on the control reefs had died. No oil. We figured out that it was clearly a disease epidemic that had swept through. At that point I said, "Darn, I know

I'm right. I know that people are somehow destroying the ecological balance in the Caribbean and so we have these diseases that kill the sea urchins [and] that are killing the corals, and we have bleaching starting to occur." In other words, the rules had changed.

IB: The Anthropocene is here?

JJ: Nobody had invented the term Anthropocene yet, but basically, in my mind, I realized that we were in what we now call the Anthropocene. We were in an era where all the rules were changed because of the overwhelming impact of *Homo sapiens*, period. . . . And that is when I shifted gears. It was in 1988 that the controls died. I'm following all this, I'm out in the water a lot, I'm running the oil-spill study, I'm the puppet master and all the people doing the fieldwork are coming back and we're talking about stuff and we're having seminars and we're learning, you know, it's exciting, we're learning—it's depressing, but we're learning.

And then I went to a meeting. And the title of the meeting was "Coral Reefs: Health, Hazards, and History." Geologists wanted to know whether the ecologists were crying wolf or if it was really a bad time. And if you didn't have history and perspective, how could you tell? At any rate, at this meeting, [the ecologists] were all standing up and talking about how this is so terrible. Finally, I couldn't take it any longer and I stood up said: "I get it, I understand it, and if there will be any coral reefs left to be affected by bleaching and acidification, it will [indeed] be catastrophic. But we might not have any coral reefs left because of overfishing and pollution—all this shit we're doing to them." And I was just pummeled with virtual rotten tomatoes and rotten eggs. "Oh, Jeremy, maybe the Caribbean is bad, but in our pristine Australia, in our pristine Indonesia," blah blah blah.

IB: What do you think about this term, *pristine?*

JJ: These are not new ideas, and historians just laugh at this debate about shifting baselines. They say, "Isn't it obvious? Look at the European landscape. Scotland used to be forest. Heather and heath is an artificial landscape created by human disturbance." I mean, it's like, "Duh!" But in the ocean, there's all this "pristine seas" bullshit.

So I wrote "Reefs since Columbus" in response.[5] After I gave the talk based on this paper to the thousand people at the 1996 meeting in Panama, you could hear a pin drop. I mean, it just totally shocked people. Because the whole point of it was that things have been going south for a long time, for a lot of reasons, and you jump on this climate change bandwagon and, yes, it's a big problem, but there's all this other stuff. So that's why I did this study.

At the end of my "Reefs since Columbus" talk in Panama, I said, "I hope this talk will do two things," and I don't remember what the second one was off the top of my head, but the first one was "that you'll never again use the word *pristine*, because nothing is pristine." This was in the days before PowerPoint, when people had slides that they brought to their talks. People took little fine India ink pens and they crossed out the word *pristine* from their talks [laughs]. The whole point of that paper is we've been doing it for a long time—and climate change is the latest thing.

IB: So where do you stand on the hope-despair pendulum?

JJ: Nancy [Knowlton—Jackson's wife] is the hope and change, and I'm less that way. And I worry sometimes about Nancy with her earth optimism being a little too rose-colored. I would say Terry Hughes's heroic effort to get the World Heritage people to keep Australia on pins and needles is an example of a committed person trying to do stuff. In our [new] Anthropocene paper,[6] by the way, we basically say that there's good news in climate change—we're actually making progress in reducing emissions. So the scenarios that call for it getting worse but then getting better are looking more and more like where the world is going to go. And that means [that] instead of looking at total catastrophe, we should be looking at the world going through a bottleneck, which is going to be really bad and really desperate, but there's light at the end of that tunnel. So we should be realists about that and do our conservation that way.

IB: What does being a realist mean in this context?

JJ: There's a lot of good news in coral reef science. I mean, corals are moving south, down western Australia, and they're moving up the Japanese coast. Even in the Caribbean, there's a lot of hope along with all the doom and gloom. But I believe very strongly that you ain't gonna get there unless you recognize the downside and you try to do something in the meantime in the context of this reality. That would also apply to Madeleine [van

Oppen] and Ruth [Gates]. They're realists, they know the problem, they think they can do something that will make a difference. And more f***ing power to them to try. [There is a] latent gender bias, which we all know, in the marine science world. It's a lot better now, but in the old days there was no doubt of the gender bias and it is a fact that there still has not been a Darwin medalist of the [International Coral Reef Society] that's a woman. I used to joke at Scripps: "What is an oceanographer? An oceanographer is a white man with a beard who wears a flannel shirt." And most women don't have beards.

IB: What do you think about restoration?

JJ: I'm a big fan of restoration—if you can clean up the place you're trying to restore. But to raise a bunch of beautiful, innocent corals and have them be healthy, young studs, and stick them out in the reef where they died in the first place is like sending them to a concentration camp. If you can't clean up the environment, if you can't make the water good, then what the hell do you think you're doing? You're either a fool or you're a sadist. And I said that to the whole Florida Keys [National Marine] Sanctuary board in a talk in Key West two years ago, in response to a very aggressive question by some coral farmer. [A] professional fisherman on the board defended me in the argument. He said, "We may not like it, but everything he said in his talk is true." Restoration makes sense in some places, if you fix the problem. Climate change isn't the only problem.

IB: Where, for example, does it make sense?

JJ: In Bermuda. The coral reefs of Bermuda were dying because of extreme overfishing, pollution from the land, [and] dredging in the harbor. Coral reef cover in Bermuda today [has] increased to about 30 percent, because they banned fish pots and they stopped the pollution from the land. It works! So if you wanted to restore a reef in Bermuda it would make perfect sense because the environment is hospitable. There are people in Curaçao and Bonaire thinking the same way.

IB: Tell me about your 2014 report on the Caribbean reefs.

JJ: My huge, three-hundred-page report on Caribbean coral reefs? I can tell you that I didn't want to do [it]. I was pushed by the person who runs the ocean program to take over something called the Global Coral Reef Monitoring Network. I was a big deal and I was retiring, so this guy asked me to do it and I said, "Okay, I'll do it on one condition: I'm not even going to pretend to do a global assessment. I'll do an assessment of the Caribbean, which everybody has written off." The World Wildlife Fund gave up on it and stopped all its work there. But there have been all these reports of places that are actually doing well, so what's going on? So I did this incredibly exhaustive study with more than thirty-five thousand quantitative surveys of coral reefs [in the Caribbean] from 1970 to the present. The world community gave us their data. I'm very proud of this.[7]

And what we found is really interesting. There's been this big decline in coral cover, which started long before coral bleaching. In fact, it was basically over before climate change started to have a real effect. The primary causes of the collapse of Caribbean corals were a combination of pollution from the land, habitat destruction, and overfishing.

IB: What was the reaction to the report?

JJ: I am hated for this report. Because my message is that Caribbean coral reefs died because of diseases [that] happened long before there was any direct climate change effect. And we've got all this proof of decrease in water quality and overfishing and tropic cascades and all the rest of it. And look: here are a few places that don't fit the mold, and the reason they don't fit the mold is clearly related to wise action. So what's wrong with this picture? What's wrong with this picture is what's right with it: that when you have a healthy fish community and you don't have all this shit coming offshore, [corals can recover]. The coral reefs of the Florida Keys, people were like, "Oh, we don't understand, it must be climate change." Bullshit. Florida Keys are the playground for five million people who live in Miami. All of south Florida is an ecocatastrophe because of human pressure and abuse. And yes, climate change is a problem on top of all that stuff, but I go back to what I said: if you could wave your magic wand and make coral bleaching and climate change go away, coral reefs in the Florida Keys would [still] be zero.

At the same time, reefs that weren't fished [have also suffered] total death and destruction. So nobody is saying climate change isn't a problem. But we don't think it's hopeless. And we think there are refuges: places where corals can go where the temperature doesn't get so high. Our paper in *Nature* says, point blank: it is really, really bad, but it's not going to be as bad as the doomsayers have been saying.[8] And there is hope, because it really looks like we're on track to stabilize the climate: business as usual is no longer supported by the evidence.

When this report came out, there were people who said, "Well, I haven't read the report but it's obviously false, and Jackson obviously is a naysayer, and he's dangerous; he's a climate change denialist." John Bruno [see my discussion of the Bruno-Cinner debate in chapter 1—IB] called me a climate change denialist on the *Huffington Post*. [How can he say that about] the person who gives talks titled "Ocean Apocalypse"?! Last year, I was in the Senate twice and at the White House once, all because of Senator Sheldon Whitehouse, who thinks that the way I talk about these things actually has resonance and could have some political effect. I did it as a labor of love. I love the Caribbean. The Caribbean nurtured me. I became a scientist, and a good one, because of my wonderful experiences in the Caribbean, and I have a passion for the place. I was on the board of the World Wildlife Fund. And when they said they were pulling out of the Caribbean, I said, "You're fools. You're going to go into this coral triangle[9] boondoggle and you're abandoning a place where people need you and where you could make a difference." And the coral triangle has turned out to be one of the biggest money-wasting boondoggles of all time.

IB: How would you summarize your most important realization?

JJ: The realization was that to understand the present, you have to understand all these things we've done—that *pristine* is a meaningless term. It's sort of like Bill McKibben's book *The End of Nature*. Amazonian forests and the most untouched places are not pristine because of carbon dioxide, period. Forests have changed. We know that. The pristine is gone—it's been gone for a long time.

IB: So how do you then justify the need to continue to try and protect?

JJ: You do it by the realization that of course you're not going to have nature in an original sense. But as a geologist and paleontologist I know that nature has always changed. The point is there's a great beauty out there [and] there's a great biodiversity. It's not the way it was, but it's great. And we can guarantee that more will exist in the future than not if we begin to take action now. It'll never be the same, but it'll be beautiful. There will be biodiversity, there'll be lots and lots of other species. Corals are going to go through an incredibly tough time, but they will survive—some of them will survive. Communities will re-form, they will readjust, and the world will go on. And you know what? We're really not much more of an agent of change than the asteroid that hit at the end of the Cretaceous. And life went on, period. It's all about not giving up, it's all about [not acting] like these idiots who are running around saying, "The sky is falling, the sky is falling!" Do you know that one-quarter of the world's electricity is already produced from renewables? The train has left the station. Businesses figured out that the smart money is in renewables. We are having a major social change. It is a revolution. Oil and coal are on their way out in spite of Donald Trump. It's a big deal and it's very positive and it's very optimistic and extraordinarily relevant to the environment.

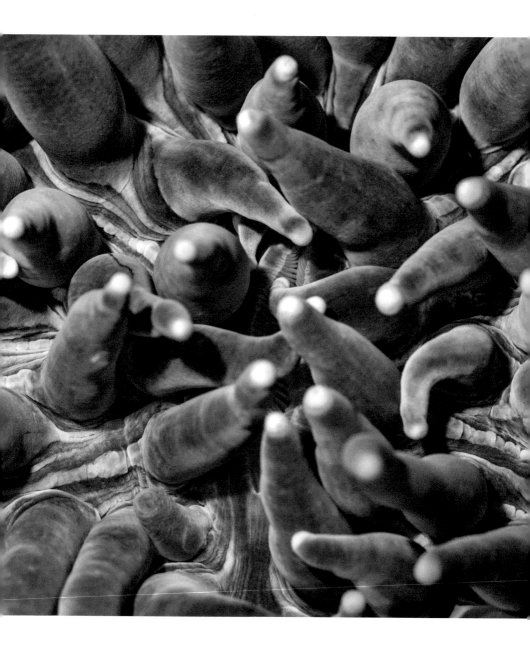

CHAPTER 3

Fragments of Hope

Nursing Corals Back to Life

Not easily, not without battling despair, I *do* live in hope.
—Scott Russell Sanders, *Hunting for Hope*[1]

STRING FIGURES

I first heard about coral nurseries while doing my fieldwork on the management of endangered toads in Puerto Rico. The biologist I interviewed, who works for Puerto Rico's environmental agency, offhandedly referred to the agency's work of restoring local corals devastated by white band disease and hurricanes. "We cultivate the surviving coral fragments until they are old enough to be reintroduced back into the reef," she told me.[2] This propagation was happening in underwater ocean nurseries, not in a lab or aquarium on land. That was a big change from the captive breeding programs I was familiar with, which occurred in ex situ settings such as zoos.[3]

But what really blew my mind was the technology used to attach the corals to the cement and, later, to the reefs: dental floss.[4] Battling disease and climate change with dental floss seemed, well, epic—like an underwater iteration of David and Goliath or, to use a darker analogy, like Sisyphus pushing his rock uphill. One year later, I visited the Puerto Rican island of Culebra to check out the dental floss arrangements at the coral nursery there. As I observed the scientists perform daily maintenance operations on the threatened *Acropora cervicornis* in the surreal underwater structure of the "horizontal line" nursery secured to the ocean's bottom, I was struck not only by the aesthetics of the

FIGURE 20. *Heliofungia actiniformis*, a long-tentacled mushroom plate coral. Courtesy of Coral Morphologic.

scene, but also by its heart-wrenching vulnerability. While the structures were simple, their designers were attempting to do something profound: to flip the trajectory of corals from despair, through action, to hope.

Afterward, I visited other coral nurseries, as well as coral farms, aquariums, labs, and restoration and artificial coral sites in the Red Sea, the Arava desert, Hawaiʻi, Florida, and, closer to home, in Boston, Massachusetts, and Buffalo, New York. My goal has been to document these projects, understand their history, and record the antagonistic responses they have spurred within the coral science community. Indeed, the project of coral restoration, although still in its infancy, has already garnered quite a few vocal opponents in the coral conservation world. This opposition was always presented to me as purely scientific, which is why the vehemence of the scientists criticizing restoration caught me off guard. The image of a restoration scientist pleading to a large crowd of coral conservation scientists, "Give us the dead and dying corals," and the very uneasy silence that followed this plea are but one memorable testament to this intensity.[5]

Because this wasn't the typical response to restoration in other ecological contexts, I was curious about the reasons for the hostile reaction to restoration by some coral scientists. Certainly, a central force behind this antagonism is the intuitive association of the ocean with pristine wilderness and, moreover, with the very last frontier of wilderness on the planet. The project of tinkering with the ocean's coral ecosystems—"coral gardening," as some have called it—therefore engenders strong resistance among many conservation scientists who were brought up on the fundamental idea that "the first principle of coral conservation is 'Don't touch the coral.'"[6] A related tension that has played out in this context is that between global-first approaches to coral conservation— which hold that restoration will fail if not preceded by global reductions in carbon dioxide (CO_2) emissions—and local-first approaches such as coral restoration, which save locally in order to "buy time" on a global scale. Finally, a less documented tension, but one worth noting, is that between in situ and ex situ conservation and, correspondingly, between the marine biologists who study corals in the wild and the hobbyists and aquarists who cultivate captive corals. Coral farmers and restoration experts in ocean nurseries are hybrids between these two worlds and are thus deemed less authentic and also less scientific than their colleagues from field biology. Despite the tensions among the various coral caretakers, coral scientists are increasingly abandoning the "conservation-only" ship and hopping aboard the "restoration-too" lifeboats.

This chapter will show that coral restoration has transformed significantly since the 1990s, when it involved the fragmentation and transplantation of the easiest-to-grow corals in the largest quantities possible, without consideration of their genetics or ecosystemic properties. The recent and rapid turn of coral science to restoration has included the development of progressively complex sets of tools, techniques, and standards across a range of geographic sites. The latest and most radical shift within coral restoration has been toward genomics. This shift encompasses the microscalar study of coral genetics and epigenetics, a heightened focus on controlled sexual reproduction, and the development of extra-resilient "super corals" through assisted evolution.

To bolster the scientific authority of coral restoration, an analogy has been drawn between corals and forests generally, and between coral restoration science and silviculture in particular. At the heels of this analogy between reefs and forests, coral restoration has also often been referred to as gardening. Naturally, the nursery has become the central locus for restoration practices. But as this chapter will show, nursing and gardening are not perceived by proponents of coral restoration as a pastime activity; rather, they are viewed as a direct and potentially effective response to the massive loss of corals and as a concrete form of action that could provide many scientists with hope. This chapter will also show that this form of active hope does not necessarily translate into optimism about coral futures.

SITES OF CORAL PROPAGATION

Corals are propagated in a wide range of sites and for a variety of reasons, of which restoration for conservation purposes is only one. Corals are propagated for public education exhibits in aquariums and zoos, for profit in the ornamental trade, for research and biomedicine, and for the mariculture industry—a brand of aquaculture developed in the late nineteenth century that involves the cultivation of marine organisms in saltwater.[7] As a tool in conservation management, cultivated corals can be transplanted back onto the reef, usually onto areas damaged by ship or storm, in order to improve coral recovery.[8] This form of management is often referred to as *restoration,* and some have suggested calling the specific, two-step process of growing coral fragments and then transplanting them *coral gardening.*[9] For the Foucauldian readers, let me offer up front that breeding corals for conservation purposes is yet another instance of biopolitics, in the sense that these processes not only protect and save valuable coral life but also attempt to literally "make live" by fostering entire populations into existence.[10]

FIGURE 21. Assaf Shaham holds an *Acropora* coral, which he farms for the bone and dental transplant industry. OkCoral farm, Arava, southern Israel. Photo by author, June 8, 2015.

Generally, underwater coral gardening projects include restoring devastated reefs using nursery cultivation and transplantation, as well as eliminating "pest" and "invasive" species such as the crown-of-thorns starfish in Australia[11] and the red lionfish in the Caribbean.[12] Restoration sites range from locations selected for their ecological, cultural, or economic significance, to mini-projects mushrooming in beach and island resorts around the world, where guests are invited to plant a coral or to sponsor a concrete coral frame with their name engraved on it.[13] Restoration projects vary not only in aims, geography, and target species, but also in technique. Corals can be grown on metal cages, ropes, nets, and PVC trees, on the seabed or in various positions in midwater; they can be propagated by both asexual and sexual methods; and they can be "improved upon" by utilizing the still experimental, and highly controversial, assisted evolution method.[14]

While Culebra's nursery propagates *Acropora* corals for restoration purposes, Assaf Shaham's OkCoral farm, located in southern Israel, aims to produce boutique *Acropora* coral (and three additional species) for bone and dental implants in the biomedical industry (figure 21). Shaham, whose expertise is in LED lighting technology and not in the life sciences, taught me that corals are

FIGURE 22. Mary Alice Coffroth's Buffalo Undersea Reef Research Culture Collection at the University at Buffalo contains approximately 750 *Symbiodinium* cultures. Coffroth explained: "I started working on *Symbiodinium* because although they played a huge role in reef ecology, relatively little was known about them and their interactions with their coral host. I am examining the potential for the symbionts to evolve and adapt to higher temperatures. We know there is much variation within a symbiont species, and if selection can act to increase the symbiont's thermal tolerance, it means we as humans have a bit more time to address the problems leading to the increase in sea surface temperatures and the crisis that threatens reefs globally" (e-mail communication). Photo by author, April 2015.

successful for knee and other bone replacements. "The body doesn't recognize [the coral replacement], but doesn't compete with it. It just stays there forever," he told me in our interview.[15] The OkCoral farm is relatively small and is located in the most ex situ location imaginable—a hot and desolate site in the midst of the Arava desert. In stark contrast, the largest coral farms in the world are situated in Indonesia's ocean waters, where corals are manufactured in high numbers for the coral trade—a topic I will briefly discuss later in this chapter.

One can also find corals and their symbiotic algae at university and research labs. Mary Alice Coffroth's Buffalo Undersea Reef Research Culture Collection at the University at Buffalo contains approximately 750 *Symbiodinium* cultures, some assembled as early as the 1970s, making it the largest ex situ collection of symbiotic coral algae for research purposes in the world (figure 22). This collection, which is a short five-minute walk from my office, investigates the conditions of coral-algae symbiosis and provides low-fee samples to research labs around the globe. Unlike corals, whose export necessitates permits under the 1975 Convention on International Trade in Endangered Species of Wild Fauna and Flora (CITES), algae do not require permits by this international treaty and so, for the most part, no questions are asked about their transportation. Coffroth's work on these algae was tucked away from the limelight until recently, when scientists figured out that bleaching events revolve around the collapse of the symbiotic algae-coral relationship. Somewhat to her chagrin, Coffroth's work has become cutting-edge almost overnight, with numerous prospective doctoral students knocking at her door.

These starkly different coral habitats illuminate an important question: What does coral life mean outside of the ocean environment? At the very least, coral nurseries and farms problematize the division between wild and captive, raising a series of complex cultural, ethical, and regulatory questions and highlighting both the distinctions and the correlations between land and ocean management.[16]

RESTORING CORAL: A BRIEF HISTORY

Until the 1990s, countries concerned with conserving their coral reefs relied heavily on a limited number of conventional strategies. Mainly, reef systems were designated as natural reserves or marine parks.[17] While this process aided corals in important ways, delimiting protected areas—especially near coastal communities—also dramatically increased tourism, thereby resulting in human trampling and sediment resuspension, which in turn caused a rapid deterioration of many reefs.[18] A burgeoning body of research has highlighted the deleterious effects of everything from cruise-ship and resort sewage run-off to less obvious pollutants such as sunscreen, nearly fourteen thousand tons of which coat coral reefs each year. Sunscreen contains oxybenzone, a phototoxicant chemical.[19] Even small doses of oxybenzone—about a drop in six-and-a-half Olympic swimming pools—can seriously harm coral larvae. Researchers found concentrations twelve times that rate in popular waters off Hawai'i and the U.S. Virgin Islands. As a result, Hawaiian lawmakers have proposed a ban on the use of such sunscreen on the islands.

Faced with the myriad convergent causes of coral damage on the local scale and with the accelerating impacts of climate change and ocean acidifica-

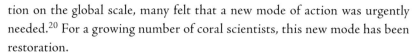

tion on the global scale, many felt that a new mode of action was urgently needed.[20] For a growing number of coral scientists, this new mode has been restoration.

The Society for Ecological Restoration defines *restoration* as "the process of assisting the recovery of an ecosystem that has been degraded, damaged, or destroyed" where natural recovery is hindered.[21] Until recently, modern marine restoration efforts have focused on three ecosystems: mangroves, seagrass, and oysters. Coral restoration is a relatively new addition to these disciplines and, some have argued, one of the least successful and most expensive of the four.[22] Initially, restoration efforts were focused on recovering a healthy reef from local and confined damage inflicted by a ship grounding accident or a severe storm. Some scientists still believe that this is the only useful function of coral restoration, and that it "is a lot less useful in cases of disease or climate change, because the causal factors remain in place and will harm the introduced corals."[23]

However, others hold that restoration in the Anthropocene entails enhancing reef resilience, a particularly important endeavor in these uncertain times. Along these lines, James Byrne of the Nature Conservancy proclaimed that "how these reefs are actively managed plays a big role in how resilient they are to warming ocean temperatures."[24] As one of the largest conservation organizations in the world, the involvement of the Nature Conservancy in coral restoration says something about how mainstream such practices have become. In 2005, the Nature Conservancy launched the Reef Resilience Program, a partnership effort that "builds the capacity of reef managers and practitioners around the world to better address the local impacts on coral reefs from climate change and other stressors." Climate change has impelled coral conservationists to act locally—on a global scale.

The verb *to act* has been central to the growing field of coral restoration and is often contrasted with "passive restoration" in the form of protection and preservation. Recent research has shown that such passive restoration through marine protection, while improving the situation for higher trophic organisms such as fish, is not as effective for the conservation of corals and other reef organisms.[25] Israeli marine biologist Baruch Rinkevich was among the first coral scientists to call for a turn to "active restoration" through the propagation of corals in ocean nurseries. For many years, he was a lone voice in proposing the application of principles already developed in terrestrial silviculture (the science of growing and tending to forests) and in advocating the use of gardening as a guiding metaphor for coral restoration.

In a 2008 article entitled "Management of Coral Reefs: We Have Gone Wrong When Neglecting Active Reef Restoration," Rinkevich expressed his frustration with the reluctance of the coral conservation community to embrace the techniques of active restoration that he had begun to champion some fourteen years earlier. Referring to marine protected areas as "a great idea with limited success," Rinkevich insisted that if humans wish to forestall accelerating coral reef destruction, it is imperative that they both employ an active approach to reef restoration that is based on silviculture practices and also begin to utilize the principles and techniques of gardening.[26] "Reef management, as executed, is failing," he wrote in a characteristically provocative manner.[27]

The recent dramatic global decline of coral reefs has pushed some coral scientists toward such practices of active restoration. The wider coral science community, too, has shifted from outright condemnation to at least partial recognition of the importance and urgency of this field. "Only three international conferences [i.e., twelve years] ago, there were just two restoration papers," Rinkevich told me when we spoke at the International Coral Reef Symposium in Hawai'i.[28] "And look what is happening now," he said, showing me the conference booklet that listed dozens of presentations organized around this theme. When walking me through his coastal coral nursery at the Anuenue Fisheries Research Center in Honolulu, coral specialist Zac Forsman told me, along the same lines, that "restoration has really changed the way folks are thinking. We have been planting forests for hundreds of years. Why can't we do something similar underwater? The process of doing so may teach us a lot. It may also buy us some time to slow down or prevent biodiversity and habitat loss."[29]

CORAL RESTORATION AND SILVICULTURE

One of the central criticisms directed toward restoration is that it is unscientific—namely, that many of the restoration projects conducted to date have been designed and implemented in an ad hoc manner, without hypothesis-based testing or long-term monitoring.[30] Marine biologist Sarah Frias-Torres has responded to this accusation by situating coral restoration within the well-established, two-hundred-year-old scientific tradition of restoration ecology. She explained in our interview that while it was developed in the context of terrestrial ecosystems, restoration ecology operates through scientific principles that can be translated to the marine environment. This, she told me, is precisely how the fields of mangrove and seagrass restoration have emerged. "When it comes to coral reef restoration, we're breaking new ground," she admitted, adding: "If we follow the scientific principles of ecological restoration—we're doing science."[31] Applying the scientific principles of the broader field of ecological restoration is thus an attempt on the part of coral restoration proponents to ground their practices in sound science.

Alongside ecological restoration, another strong scientific anchor for coral restoration has been silviculture, or the science of forestation. Drawing a straight line from tree management to coral management is not as simple as it sounds: unlike trees, corals are animals, and unlike most trees, they live in the ocean. Rinkevich explained that silviculture principles, concepts, and theories naturally lend themselves to coral reef restoration because restoration ecology has always been biased toward botanical issues.[32] Applying botanical principles to corals is easier because of their similarities to plants—their asexual reproduction, their ability to phosynthesize, and their ecosystem function as "rainforests of the ocean."[33]

Once the analogy between forests and reefs was drawn, it was only a matter of time until the broader relevance of this analogy to coral management would be realized. Accordingly, Rinkevich then cited that "around 95 percent of the forests in Canada and the United States are man-made. More than 99 percent of forests in Europe are man-made."[34] He was referring to both the planting of commercial forests as well as the intensive management of noncommercial ones. "People understood early on that they should restore the forest by transplanting trees," he explained.[35] For him, this raises the obvious question: "Can we, or should we, achieve the same state-of-the-art in reef management? Can reef restoration acts mirror those of silviculture?"[36]

Similar to trees in terrestrial forests, corals, too, generate three-dimensional structures. Such "marine animal forests," as certain marine scientists have referred to them, are increasingly understood as providing ecosystem services like food, protection, and nursery to their associated fauna and as playing an important role in the "local hydrodynamic and biogeochemical cycles" near the seafloor, including acting as carbon sinks.[37]

CORAL NURSERIES

Following in the footsteps of the terrestrial tree nursery, the coral nursery has emerged as the central site of coral restoration. The first coral nurseries for restoration purposes were built in the Caribbean in the 1990s, as the full extent of coral degradation was realized. Rinkevich built his first coral nursery in 1995 in Eilat, Israel. A one-year experiment at the nursery ended, instead, after seven months. Rinkevich explained: "The reason that we terminated the experiment was that the corals grew so fast—three times faster than the natural corals."[38] Corals are territorial animals and "superb fighters,"[39] he told me. He was thus concerned that instead of propagating and recovering, their amazing growth rates would result in their killing each other. Far from being passive recipients of restoration, then, corals emerged in Rinkevich's experiment as active participants, deploying their own resistance tactics, which frustrated those of the experts. As a side note, I'd like to point out that thinking of territorialism in corals can be jarring to those of us accustomed to the more cited collaborative and mutualistic features of corals that have provided inspiration to so many thinkers.

Drawing on research conducted during the 1970s[40] and 1980s,[41] Rinkevich continued to demonstrate the feasibility and success of coral transplantation. "These experiments," he wrote, "revealed that 80 percent of coral fragments, imported from nearby reef and transplanted onto dead reef frameworks, were still alive after three years."[42] Based on this research, he devised a process whereby asexual coral recruits and fragments were clipped to concrete plates and placed "in favorable shallow-water sites within the reef until the colonies reach[ed] expected size and the coral fragments regenerate[d], forming the structure of a colony specific to the studied species."[43] Later, techniques for propagating corals in midwater floating nurseries[44] and in rope nurseries[45] were developed, allowing for quicker and more efficient coral growth (figure 23).

Designed by Rinkevich and located near Eilat's northern beach in the Red Sea, the world's first midwater floating nursery propagated eight thousand coral fragments. Created within the last twenty years, this nursery was the first attempt of its kind to raise thousands of coral colonies at a single site. An even newer technique for coral propagation, the midwater rope nursery, was installed in the Philippines in 2006.[46] Rinkevich explained: "You take a rope of twenty meters, you open the twist of the threads, you insert the coral fragment, and then you leave it at five to seven meters above the bottom. Within

FIGURE 23. Elad Rachmilowitch (then a graduate student) adds new coral nubbins (minute coral fragments of one to several polyps) to the midwater bed nursery in Eilat, Israel, after the transplantation of grown corals. Each nubbin is glued to a plastic pin, and all the pins are inserted into quadrates of framed fine mesh. The nursery is situated at six meters' depth and twenty meters from the substrate. Photo by S. Shafir, 2012. Courtesy of Baruch Rinkevich.

one year you can transplant the entire rope either on empty rocks or on the sand."[47] The central advantage of the rope nursery technique is its ease of transportation: the ropes can be lowered to the seabed during storm events and can also be outplanted at the target site for restoration, along with the diverse life-forms that have developed around them, without causing too much stress to the corals. In Thailand, a team of six people successfully transplanted five thousand such rope-coral assemblages in one day, lowering the cost of restoration to sixteen cents per coral colony.[48]

Reducing the cost and increasing the scale of production are important preconditions for making restoration into a feasible conservation enterprise in the Anthropocene. Still, critics of coral restoration repeatedly refer to its high price. According to one study, gardening one square meter of corals can cost up to $500. Others criticize the limited geographic scale of coral restoration. A recent fact sheet from a Florida nursery sets a goal of restoring one thousand acres, or 1.5 square miles, of reef in the Florida Keys within the next ten years; this, when the entire Florida Reef tract is six hundred square miles and the Great Barrier Reef—the world's largest—is more than two hundred times that size.[49] The scale of restoration, the critics imply, is simply inadequate. "I agree that restoration is better than doing nothing," environmental sociologist Joshua Cinner (recall his "bright spots") told the press. "But your options aren't doing nothing or doing restoration. The millions of dollars spent on restoration could be spent improving water quality, managing fisheries, and reducing impacts of tourists. If [done well], these all can have tangible impacts on improving coral reefs."[50]

Proponents of restoration have responded differently to the criticisms. While some have highlighted that success is not necessarily a matter of scale,[51] others have maintained that restoration on an appropriate scale is not off the table. Rinkevich argued, for example, that while traditional nursery practices have focused on growing hundreds of corals for transplantation, it is possible to create a mega-scale restoration operation that would include "raising hundreds of thousands of corals and then transplanting them."[52] Rinkevich insists that employing such ambitious approaches would enhance the ability of coral reef organisms to respond to climate change. In his words: "The continuous degradation of coral reef ecosystems on a global level, the disheartening expectations of a gloomy future for reefs' statuses, the failure of traditional conservation acts to revive most of the degrading reefs and the understanding that it is unlikely that future reefs will return to historic conditions, all call for novel management approaches."[53] This statement illustrates that even among the fiercest promoters of restoration through coral gardening, gardening is not performed for its own sake; it is, instead, a way to do something meaningful in the face of degradation. Other proponents of restoration have highlighted, accordingly, that "regrowing corals won't work in areas in which the basic reason they died off in the first place has not been fixed." As Stanford coral scientist Steve Palumbi put it, there's "no point in trying to put a vegetable garden on a landfill."[54] If temperatures continue to climb, many scientists have stressed, no part of the ocean will be amenable to nurseries and replanting.

Alongside Baruch Rinkevich, Ken Nedimyer is another paradigm shifter in the world of coral restoration. Situated in the Florida Keys, Nedimyer's Coral Restoration Foundation, a nonprofit he established in 2007, consists of five nurseries that together comprise the largest coral nursery effort in the world, with half a million corals propagated for transplantation as of 2017.[55] Early in the process, Nedimyer shifted much of his elkhorn and staghorn cultures to hanging PVC-pipe "trees" suspended in the water column "like a mobile that you hang over a child's crib."[56] These structures, which resemble terrestrial trees, are a physical reminder of the strong reliance of coral restoration on silviculture practices. Nedimyer also spearheaded efforts to establish sister PVC tree nurseries and to train dozens of groups to do the same across the Florida Keys and the Caribbean. Initially focused on *Acropora cervicornis*, or staghorn coral, Nedimyer recently added slower-growing corals such as *Acropora palmata*, or elkhorn coral, to his nursery stock, and is gradually adding other coral species.[57]

I would like to circle back to the realization that the silviculture model has enabled the emergence of coral restoration as a science. Twenty years after designing the first large-scale nursery, and despite his long-standing reliance on terrestrial forest management for guidance, Rinkevich—the father of coral silviculture—has come to recognize the limits of this comparison. He recently reflected on one major difference between tree and coral nurseries:

> Now we know that coral nurseries are completely different from tree nurseries because tree nurseries are sterile: there are a lot of insecticides and other compounds that [are applied to keep] the trees free of pests and other organisms. But the coral nursery is open to the environment. [So] when we transplant the coral from the nursery, we can transplant the coral with the fish and the invertebrates that are found in and between the branches. It is a completely different idea: with corals, we are raising a habitat, not just a single organism.[58]

Rinkevich's move away from silviculture arguably signals that coral restoration is now ready to stand on its own two feet. As part of its maturation, this field is becoming more integrated, more coordinated, and more standardized, attracting new adherents and giving hope to many coral scientists who were, until recently, quite skeptical about the scientific underpinnings of this practice. In the face of the accelerated decline of reef-building corals, such coral scientists, especially those of the younger generation, are increasingly playing a more active role by refocusing their work on resilience and renewal.

In our interview, Zac Forsman told me about his own changing relationship to restoration, explaining why he has eventually decided to dedicate much of his time to the new coastal coral nursery at the Anuenue Fisheries Research Center. Restoration "was the Wild West only a few years ago," he reflected. But today, "there is a clearer and urgent need and it has become a central management tool."[59] This is especially the case in places like the Caribbean, he told me, where endangered coral species have dwindled so much that they can no longer reproduce effectively. The situation was spiraling downhill until "coral nurseries have helped turn things around for some reefs."[60]

Innovative restoration techniques are also mushrooming in other parts of the world. Austin Bowden-Kerby, director of the Fiji-based nonprofit Corals for Conservation, is famous in the field for his central role in the documentary *The Coral Gardener*. His model for restoration in Fiji is different from Nedimyer's in Florida. Bowden-Kerby collects coral fragments from sites where they cannot grow for lack of space, transplants them onto raised platforms with healthy living conditions, and finally transplants them back onto the degraded reef. Over the years, he has trained local fishermen to become "coral gardeners" like himself. These locals are hired by resorts to protect the reefs, which are additionally declared "no fishing" zones.[61] "The coral captures people; it captures their heart," Bowden-Kerby reflected in one of the film's scenes. "[But] once you have people's heart, what are you going to do with it? You're going to get them to work together to try and save the planet."[62] Hope is not enough, according to Bowden-Kerby. To make a difference, hope must be coupled with practice. Macy and Johnstone wrote along these lines: "We don't wait until we are sure of success. We don't limit our choices to outcomes that seem likely. Instead, we focus on what we truly, deeply, long for, and then we proceed to take determined steps in that direction."[63] Put differently, while passive hope "is about waiting for external agencies to bring about what we desire," active hope "is about becoming active participants in bringing about what we hope for."[64]

In the Caribbean, where the decline of coral ecosystems and the resultant emergence of coral nurseries occurred long before climate change was widely acknowledged, NOAA has recently increased its efforts to coordinate between the multiple nurseries and to design standards and procedures for their operation.[65] Local organizations are also establishing standards for specific contexts. For example, the 2014 *Best Practices Manual for Caribbean Acropora Restoration* provides a set of standards for coral nurseries, focusing on technical issues such as how to build hurricane-resistant steel-bar tables to support the nurseries and how to secure a rope culture of corals to these tables.[66] Such standards tap into the network of coral nursery practitioners, who until recently were exchanging information about coral husbandry and survival mainly through informal channels, which is partly why their work has not been recognized as serious science by many marine biologists. Tom Moore,

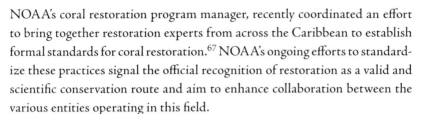

NOAA's coral restoration program manager, recently coordinated an effort to bring together restoration experts from across the Caribbean to establish formal standards for coral restoration.[67] NOAA's ongoing efforts to standardize these practices signal the official recognition of restoration as a valid and scientific conservation route and aim to enhance collaboration between the various entities operating in this field.

Just how much, and how quickly, times are changing for coral restoration was evident from my conversations with NOAA research ecologist Margaret Miller.[68] As mentioned earlier in this book, although Miller was quite skeptical about coral restoration when we first spoke in 2015, her views had fundamentally changed less than a year later. "We are in an era of proactive conservation," she told me in 2016, "a time when much more meddling is required than we were ever willing to do before." For corals in particular, she said, "the wild no longer exists . . . because the environment is changing too radically and too fast."[69] Not only the environment, but also coral administrators, managers, and scientists are needing to adapt quickly to the changing times and climates.

CORAL FARMS

Early in my interview process, I was taught to distinguish between coral nurseries and coral farms. A nursery is an intensively managed coral restoration site intended for conservation purposes, I was told, whereas a farm exists for the commercial production of corals. The underlying normative assumptions of the coral scientists I spoke with were clear: one must distinguish the good (putting more corals into the ocean in order to save them) from the bad (taking corals out of the ocean in order to reap profit). Yet I soon discovered that the nursery/farm distinction was not as straightforward as it was made out to be. A messy reality has surfaced, which includes a complex web of interrelations among coral restoration scientists, aquarists working in zoos and aquariums, and hobbyists in the mariculture trade.

Trade regulations enacted through CITES aim to ensure that international trade in specimens of threatened wild animals and plants does not negatively affect their survival. Despite the listing of certain coral species under CITES Appendix II, however, the coral trade is a booming industry. Browsing the Internet, one can find every coral species under the sun, including those designated as threatened under the law, offered for sale in a variety of appealing shapes and colors and available with overnight shipping. "When eBay first got started, one of the first things that got put on there was aquarium animals, fish and coral," one aquarist told me.[70] Indeed, a cursory search on eBay provided 5,976 results for coral and live rock, including frag kits, cured reef plugs, and special coral glue.[71] This is just the tip of the iceberg, I was told; coral clubs have mushroomed across the country, accompanied by a tight social network on Facebook.[72]

A terrestrial architect by training, Walt Smith has been commercially harvesting fish and coral in Fiji since 1972. In our interview, Smith acknowledged that, like other coral farmers, he has a bad reputation among coral scientists: "There are some serious groups out there that would really like to see the aquarium industry just disappear." He attributed this foul reputation to sheer ignorance. While agreeing that there are some "bad operators" in the field of coral farming—Indonesia having the worst of the lot, in his view—Smith rejected the treatment of the coral farming trade as altogether bad. "Let's face it," he told me,

> the aquarium industry creates more awareness about coral reefs than any other industry or any other science possibly could. Show me a parent who doesn't like to take their kids to a public aquarium and hold them up to look at the fish and the coral reefs. It's an amazing thing. Very few of those people would ever get a chance to come down to the South Pacific or go out to the middle of the Caribbean and see this stuff for real in nature.[73]

Although Smith would prefer for the majority of coral harvesting to take place at in situ farms rather than in the wild, he nonetheless argues that even wild harvesting is, fundamentally, a sustainable activity. "The aquarium trade is very selective," he explained. "We don't just go down and clear-cut like they do in forestry. Our harvest is 0.001 percent of any given reef source." Given the limited number of aquariums in the world, he added, the industry is relatively small. Lastly, Smith argued that there are groups that do far more damage to the reef than those that harvest it for the aquarium trade. In his words,

> In the Philippines and Indonesia, it is very widely practiced to go out and catch fish with the use of cyanide. They put cyanide in squirt bottles, like ketchup bottles, and go down to the reef and squirt the cyanide all over the reef. The fish hide in the reef, so now all these fish come running out, gasping for their last breath, they fall in the sand, and [the fishermen] pick up whatever is still alive, which is about 10 percent of what they just squirted with cyanide. The fish are doomed to die because their digestive tract is destroyed, and what they leave behind is a completely dead coral reef.

FIGURE 24. A varied assortment of *Acropora* species at the Walt Smith International coral farm in Fiji, July 2010. The structure beneath the corals is made of steel rebar and welded wire mesh. The concrete pegs (shaped like a large golf tee) are used to mount the coral frags, and the stem fits into the mesh to hold the corals in place. The table stands about sixty centimeters from the substrate. Courtesy of Walt Smith.

Smith clarified that "the coral harvesters themselves are not guilty of any unsustainable practice. But the aquarium trade itself is—because of the use of cyanide to catch fish. The reefs in the Philippines have been hugely destroyed by cyanide. And those of Indonesia as well."[74]

Alongside his convictions about how *not* to do restoration, Smith also offers a scheme for how restoration ought to be done. He is a strong believer in the importance of small coral farms for the economic and cultural survival of local communities, both human and coral (figure 24). To advance such arrangements, Smith founded the nonprofit Aquaculture Development for the Environment (ADE). This organization aims

> to establish a reef restoration and research project in Fiji that would train local villagers in the technique to replant coral fragments taken from donor brood stock and record the success of increased habitat in at least 45 different sites, at the rate

of 15 villages per year, producing 12,000 pieces of new coral in each site annually by the year 2019.[75]

The idea is that ADE will buy a percentage of these reefs every year to make the project profitable. "At the end of the day," Smith told me, "these villages will [become] the largest coral reef restoration project in the world. There is nothing else that even comes close to this."[76] Smith's farm-nursery hybrid encourages the sustainable harvesting and protection of reefs by local communities.

The complex coral trade system described by Smith flips on its head the widely shared assumption that taking corals from the wild is bad but growing them in mariculture farms is good. Today, the vast majority of legal and illegal maricultured corals originate in Indonesia, where their production is massively industrialized. The legal propagation of corals in Indonesia is regulated through CITES quotas,[77] which set up a distinction between wild and farm-raised corals. "Indonesia has an industry protocol: all the corals are tagged with a catalogue number and have an audit trail for their farms," Michael Tlusty and Andy Rhyne of the New England Aquarium explained to me after we toured the aquarium. "Now this sounds great, because you're not taking from the wild. But instead of growing corals in that remote village where people's only livelihood is fishing, now they don't do that anymore. So you've both displaced local villagers and you've industrialized the coral operation."[78] They therefore emphasized that "we think that if we're buying coral from our neighbor's basement we're saving a wild reef. But the truth of the matter is that if we buy corals from local fishermen, we've saved the reef because that person is now going to value and protect it."[79]

Colin Foord, codirector of Miami-based multimedia group Coral Morphologic, strongly agrees with Smith, Tlusty, and Rhyne about the importance of the local coral farm, referring to local incentives for ornamental coral mariculture in Indonesia and Fiji as "revolutionary." He told me,

> In my mind this is easily the most exciting thing happening in the world of coral [today]. . . . It is also the most sustainable form of aquaculture in the planet, connecting the local people to understanding the need to keep their coastal waters clean for healthy corals. Too often the scientists and nonprofit leaders [who] dictate what a developing island nation should or shouldn't do are all white and Western [and] have no real connection to the culture of the people who have been depending on these reefs for generations. But without the economic incentive, there will be no protection of coral reefs.[80]

The coral farm, which turns corals into valuable commodities outside of their ocean environment, is one way in which corals are assessed in economic terms. Increasingly, a price tag is placed on corals in other contexts as well, and even as an integral part of their conservation and restoration.[81] Indeed, some conservation scientists have argued that corals would be better protected if conservation projects were designed to be more profitable.[82]

Although this is not my focus here, I will say that such "win-win" promises of both conservation and economic growth through the commodification of nature warrant careful examination.[83] The "neoliberalization"[84] of science and the related commodification of nature[85] have been criticized extensively in the academic literature. In Marxist-inspired geography scholarship in particular, the bulk of the extensive recent research into neoliberalism and the nonhuman world has been decidedly unsympathetic to the project of "market rule."[86]

ENGINEERING CORAL ECOSYSTEMS

I have digressed somewhat into the realm of coral farming to show that the boundaries between coral nurseries and farms—and, correspondingly, those between restoration and capitalist production—are not as clear as they are made out to be. The rest of the chapter will shift from the nursery and farm to focus on the laboratory, in order to discuss the expansion of restoration practices into the domains of coral genetics and reproduction.

Not only the practices and techniques, but also the very definition of restoration have undergone major changes lately. The definition I quoted earlier from the Society for Ecological Restoration focuses on "assisting" the recovery of damaged ecosystems, with the understanding that the restored ecosystem would then be self-sustainable. Now, however, certain coral scientists are suggesting that we consider designing novel,[87] or at least hybrid,[88] ecosystems. These suggestions raise important questions: How novel is novel, and how far is one willing to go?

In a 2017 publication entitled "Shifting Paradigms in Restoration of the World's Coral Reefs," Australian-based coral geneticist Madeleine van Oppen and her colleagues suggested applying the term *intervention ecology*, originally proposed for terrestrial systems, to coral reef management.[89] In their words,

> If the historical ecosystem state is no longer attainable through natural recovery processes or through human intervention, either "hybrid" (those retaining some

original characteristics as well as novel elements) or "novel" (those that differ in composition and/or function from present and past systems) ecosystems are two possible alternative restoration objectives that have been considered in terrestrial restoration initiatives.[90]

Other coral scientists have suggested a focus on conservation actions that "buy time." This phrase has been used repeatedly by my interviewees, in what seemed like a strange disregard for the consumerist association of the word *buy*, especially as it refers to time, which seems so inherently unbuyable. In our interview, Margaret Miller of NOAA offered a simpler meaning for this phrase. For her, buying time means "maintaining some minimal population levels, basic levels of reproduction and genotypic diversity within the species, [until] we can hopefully, over time, get a handle on global warming, acidification, and disease—all these factors that are impairing natural reproduction and causing mass mortality events."[91]

According to Baruch Rinkevich, however, such acts of buying time will not suffice for the reefs of today to transition into the reefs of tomorrow. To have a chance of coral survival into the future, he argues, propagating monoculture corals in individual coral nurseries, active and large as they may be, will not be enough. He has thus been promoting a broad-scale "ecosystem engineering" of coral reefs. In his words, "the 'gardening' tenet, equipped with the tools developed for coral farming and transplantation, may serve as a unique platform for employing novel environmental engineering tactics."[92] For Rinkevich, such a novel engineering platform is far from being a choice; it is, instead, a necessity in today's rapidly changing climates.[93]

Specifically, Rinkevich's "reefs of the future" include using certain reefs as "Noah's Arks" or "climate refugia" to repopulate corals of the future, transforming midwater coral nurseries into floating ecosystems that carry within them oases of life as "a biodiversity management instrument," using nurseries as sexual-reproduction "hubs" for coral larvae, and establishing new connectivity routes between fragmented reef zones.[94] His most recent project involves engineering the ocean's substrate with tiles covered in coral so as to replace hard turf with coral-friendly materials. "Tough algae means no settlement," he told me. "By changing the substrate, it becomes artificial. [But] we don't care. We call it a facelift."[95] Some scientists have used the engineering idea to further suggest that, in worst-case scenarios, artificial reef structures be used as replacements for natural habitats. "Although engineered habitats would not protect the underlying integrity and diversity of coral reef systems, they could become essential to maintain some reef-related ecosystem goods and services," these scientists have offered.[96]

Despite his recognition that we now live in the Anthropocene, and the implications of this recognition for justifying radical intervention in coral ecoystems—including massive engineering projects that "facelift" nature—Rinkevich still believes that there are limits to what coral scientists should do. He is especially critical of the project of assisted evolution and the making of super corals. "It is one thing to select for the most resilient alleles within an existing natural population," he explained, "and it is a very different thing to manufacture and introduce more resilient corals into the natural ecosystem."[97] Evidently, even within the restoration community, which is generally more accepting of interventions in nature, worldviews vary considerably.

My outline of coral engineering approaches would not be complete without providing a few historical anecdotes. Wild ideas for fixing corals through reef engineering began to surface as early as the 1980s, for instance when Wolf Hilbertz and Thomas J. Goreau developed an "electric reef" technology called Biorock.[98] This is how it was supposed to work: two electrodes served by solar cells or wind-driven generators would supply a low-voltage direct current, which would cause minerals in the seawater to precipitate onto the steel Biorock frames; once the limestone accretion established itself, pieces of coral would then attach to the structures.[99] Ultimately, the Biorock technology for producing coral-friendly surfaces turned out to be less practical than its designers had hoped.[100]

Through the years, other geoengineering proposals for saving coral reefs have also been considered, including placing shades on floating sails to protect corals from the heat[101] and additional forms of solar radiation management.[102] Mark Eakin, coordinator of NOAA's Coral Reef Watch, told me about experimental solutions such as pumping cool water on the corals or inserting aerosol into the atmosphere.[103] Eakin is a supporter of such geoengineering fixes. In fact, he believes that much more should be mandated and regulated by the federal government and that funneling more resources in this direction is crucial. As he put it: "Why can we put tens of millions of dollars down into protecting a roadway we built in the first place and can easily rebuild, but we don't have the resources to protect a natural resource that we have no way to rebuild?"[104] The largest regulatory and technological effort, however, should be "to reverse the warming that we're seeing now," he told me. Implied in this statement is that if humans would only put their minds and technology to it, they could fix the world's conservation problems. This is a controversial proposition, and one that many other coral scientists have been wary of, especially given its underlying assumption that unpleasant lifestyle changes can be avoided through technological advances.

THE GENETIC TURN IN CORAL RESTORATION

Until recently, all coral restoration projects utilized asexual coral reproduction. Corals reproduce asexually by either budding or fragmentation. In budding, new polyps separate from their parent polyps to form new colonies, whereas in fragmentation an entire colony (rather than just one polyp) branches off to form a new colony. This may happen during a storm or a boat grounding, or as part of deliberate restoration actions.[105] Because asexual recruits do not enhance the genetic diversity within a reef, Rinkevich realized already in the mid-1990s that "the use of sexual recruits is more favorable, although it is a longer and more complex process."[106] A novel body of research is now emerging that is focused on improving the success rates of transplanted sexual recruits,[107] and nurseries and scientific research projects are amending their practices to include an enhanced focus on genomics and sexual reproduction. I will discuss two such experiments here.

Fragments of Hope is a local organization in Belize that establishes and manages coral nurseries. The organization's director, Lisa Carne, successfully adapted some of the nursery techniques developed by Austin Bowden-Kerby in Fiji. In a joint paper, they detailed their restoration efforts at Laughing Bird Caye, which lost all of its threatened acroporid corals to two major hurricanes, one major bleaching event, and multiple disease outbreaks. To restore what has legally been designated a "no-take" national park and a UNESCO World Heritage Site, their strategy has been "to create genetically diverse coral nurseries and reef restoration sites that hopefully also incorporate bleaching- and disease-resistant parent stock, gathering together corals from scattered remnants and growing them into sizable populations to increase the chances of fertilization and genetic recombination during spawning."[108]

Bowden-Kerby and Carne also designed what they call "gene bank nurseries" for the three threatened acroporid species *A. cervicornis*, *A. palmata*, and *A. prolifera*. Across six established nursery sites, they used three nursery methods: mesh A-frames, suspended ropes, and cement disks affixed to mesh trays. Overall, they planted 354 corals from seventeen genotypes. These colonies were trimmed twice to produce fragments for the restoration sites. Approximately four thousand second-generation corals were eventually planted on reef patches within Laughing Bird Caye National Park. A genetically diverse *Acropora* population containing potentially thermally tolerant genotypes has thus been reestablished. The scientists relayed their hope that this population will facilitate the restoration of sexual processes for natural recovery.[109]

Steve Palumbi's work provides a second example of the shift of current coral research and restoration practices toward genomics and sexual reproduction. Based at Stanford University, Palumbi is both an ecologist and a geneticist. As he sees it, "It all boils down to fixing the CO_2 addiction that we have. But in the meantime, our job is to save as many coral reefs as possible."[110] To do this, he asks a simple question: "Where are the corals that can survive longest in the face of future climate change?" Determined to find the corals of the world who are most resistant to higher ocean temperatures, Palumbi stumbled upon a unique environment. "We happen to work in one small area in American Samoa [that] heats up during the day at low tide. It's a coral bathtub and sometimes it's so hot! But the corals love it. Why don't they die? They're supposed to die in such temperatures." To figure out the answer, Palumbi's lab concentrated on the four most abundant of the thirty coral species thriving in this pool, which can reach 35°C during midday low tides in the summer (the rest of the population, which experiences lower variability, is rarely exposed to water temperatures higher than 32°C). "We're building a picture of what it is they do to survive," Palumbi told me. As part of this attempt, he transcribed the full genome sequences for 120 individual colonies and gathered information about the symbionts and the habitats they live in. His lab is currently establishing a coral nursery, where they will cultivate the most resilient, heat-tolerant corals they find and then replenish the natural reefs with them.[111]

Palumbi's work demonstrates two mechanisms by which corals adjust to increased temperature. The first, acclimation, is a short-term process whereby individual organisms change their physiology to become more heat tolerant. The other, adaptation, is the longer-term process of natural selection whereby certain individuals pass on the genetic traits that make them more successful in warmer environments.[112] With corals, as with many other organisms, the combination of the two factors is more powerful than either on its own (Palumbi offered the following analogy: a person who has both dark skin and a tan is less likely to sunburn than a person who has only one or the other).[113]

He emphasized that the most urgent actions need to occur at the local level, while at the same time keeping an eye on the global threats of climate change. In his words,

> The best thing you can do for a reef is protect it from other things that climate change is not affecting [but] that still kill corals: sedimentation, overfishing, habitat destruction, pollution. We can change all those things right now. And they cumulatively kill more corals than climate change does, right now. And in the meantime, we should fix the underlying problem [of] CO_2 emissions.[114]

Palumbi's work thus highlights the recent turn of coral restoration toward genomics. In the past, restoration was mostly executed through asexual propagation, whereas "the future is restoring with larvae that are created with sexual reproduction,"[115] which requires a sensitivity toward the coral's genomic makeup.

CORAL SPAWNING

Most symbiotic coral species are hermaphrodites—that is, their polyps are both male and female. The remainder have separate male and female colonies or, in solitary species, separately sexed individuals.[116] The pivotal sexual reproductive event for corals, and one that never ceases to amaze coral scientists and laypersons alike, is coral spawning. For the past couple years, Nedimyer's large *Acropora* nurseries at Key Largo have attracted a growing number of researchers and aquarists from a variety of institutions. The central attraction—coral spawning—takes place once a year, three nights after the August full moon. At this time, entire colonies of the threatened *Acropora* species simultaneously release their eggs and sperm (their gametes) into the water column. The sperm then fertilize the eggs, and the resulting embryos develop into coral larvae (or planulae), one percent of which will later settle on the ocean floor and survive the first few months of their life.[117] Researchers work around the clock to collect the gametes, trying to figure out how to enhance their sexual reproduction rate.

Linda Penfold is a reproductive scientist at the South-East Zoo Alliance for Reproduction and Conservation (SEZARC), a nonprofit group that assists zoos and aquariums with fertility problems in their endangered species.[118] She has already assisted in the fertilization of a long list of such species, including northern white rhinos, sand tiger sharks, and African elephants. "My job is to help zoos and aquariums breed their species so we don't have to take any out of the wild," she told me in our interview.[119] She noted, additionally, that "the aquariums are about twenty years behind the zoo industry with [regard to] reproduction. Historically, they have always taken their animals out of the wild. But they are realizing that it's not sustainable." Recently, the threatened *Acropora* coral was added to Penfold's fertilization list.

Because much of the knowledge on captive coral husbandry was developed by hobbyists and aquarists rather than by biologists, the former have played a major role in the recent shift of restoration into the realm of coral husbandry generally, and into that of sexual breeding in particular. Yet some of the hobbyists and aquarists have been frustrated with the lack of recognition by the coral scientists of their contribution, which is often depicted as unscientific. In the words of Colin Foord of Coral Morphologic:

> In my opinion, the real coral whisperers were the pioneers who cracked the code of coral growth and asexual reproduction in closed systems. The people who deserve the most credit for understanding the living biology of corals are largely passionate amateur hobbyist aquarists who tinkered away in their basements and garages in the 1980s and worked out the chemical [and] physical needs of corals in a closed system. . . . Without them, there would still likely be no live coral exhibits at any public aquarium in the U.S. or Europe.[120]

He continued,

> I would wager that there are more high school kids who are capably growing corals in their bedrooms than there are Ph.D. coral biologists who would even know where to start. The difference in the relationship with coral between an old-school, gray-haired coral biologist and a coral hobbyist is that the coral [biologists have] traditionally removed themselves . . . from the well-being of the coral. . . . But a hobbyist is emotionally, financially, and empathically connected to their coral. They live together 24/7.

Foord's narrative highlights that scientists are not the coral's exclusive caregivers, though many of them may feel that they are. In his world, hobbyists are the true coral whisperers and experience more intimacy with their subjects than the scientists. It may be useful to think of Foord's position as a variation on Michel Foucault's concept "the great battle of pastorship"— a power struggle over who cares more and better for their sheep.[121] Of course, this doesn't have to be a battle: in addition to the biologists, indigenous communities, hobbyists, divers, and aquarists all strive to know and care for corals.

More broadly, during my research for this project I have usually found that aquarists and hobbyists, particularly those of the younger generation, are more hopeful about corals than their scientific colleagues. "There is definitely a generational divide," Foord told me. "The oldest men tend to be the doomiest and gloomiest on the future of coral reefs, perhaps because they got to experience the reefs in their prime pre-1980s. But conversely, they are also the most naive about the actual living biology of these fluorescent fleshy creatures, and our potential to grow them."[122]

SECORE

Alongside the hobbyists, who grow corals in their homes, aquarists, who grow them in public aquariums, have also played an important role in shifting coral restoration to becoming more husbandry-focused. SECORE (SExual COral REproduction) is a prime example of the prominent role that zoo and aquarium professionals are coming to perform in the transition of coral restoration toward sexual reproduction, and toward genomics more broadly. SECORE was established in 2002 by Dirk Petersen of the Rotterdam Zoo, who still acts as its director. Iliana Baums is a coral geneticist who serves on the board of SECORE. She told me that by placing large, condom-like sleeves over corals during spawning events (figure 25), SECORE was able to successfully collect gametes, conduct ex situ fertilization, and rear larvae across the Caribbean. The tricky part has been to get the larvae to settle back in the ocean. In her words, "SECORE is trying to figure out at what point in this life expectancy it is best to take the settlers and put them back out on the reefs so you get the best survival rate for the effort that you put in."[123]

Alongside partnerships with coral scientists like Baums, SECORE also initiates collaborations with aquariums and zoos. Petersen explained that zoos "have a lot of expertise looking at husbandry" and that "this expertise is very useful for restoration work."[124] In addition to husbandry-related expertise, Petersen believes that zoos also work better at communicating with and educating the public, a central tenet of coral conservation. In his words, "Very often, scientists don't have a good connection with the public. They work in the laboratory or in the field, and nobody knows what they are doing or why they are doing it. Our experience with the zoo and aquarium network [is that] we reach a lot of people. This is really important."[125] As this book has

FIGURE 25. Iliana Baums of the Department of Biology at the Pennsylvania State University described this photo to me: "I am placing a homemade coral condom on an elkhorn colony. We use these nets to collect egg and sperm from the elkhorn corals. The collected gametes are used for experiments like genome sequencing, and to breed elkhorn larvae with the goal of reseeding the reefs. The photo was taken during a joint field operation with SECORE International in Puerto Rico in July 2009" (e-mail communication, June 16, 2017). Courtesy of Iliana Baums.

repeatedly shown, working across the traditional divides is becoming a necessity in times of emergency. Recognizing the dire state of corals, field scientists and zoo experts in particular have been initiating cross-institutional collaborations that attempt to use the relative strength of each group for the common goal of saving corals.

In April 2017, SECORE International launched a joint project with the California Academy of Sciences and the Nature Conservancy for "global coral reef restoration." Starting in the Caribbean, this project aims to seed reefs with sexually reproduced coral offspring, thereby helping to maintain the corals' genetic diversity and maximizing their ability to adapt to future conditions.[126] Petersen explained:

For pure aquarium or zoo husbandry, it's not necessary to look at genotypic diversity. Clones or non-clones, it doesn't matter. [But] if you look at our restoration work, we try to cross-fertilize as many individuals as we can. You're not talking about a rhino that might be shot by somebody; you're talking about an ecosystem under stress impacts. It's a very different story. Of course, you have thousands of coral colonies out there, but it can take [only] a few events and it's all over.[127]

Over the past several decades, accredited zoos have become experts in the captive management of endangered populations generally, and in their reproductive management in particular. Specifically, zoos have spearheaded the development of models for managing small populations in the wild. In addition, their captive-breeding expertise has become crucial for many taxa, such as snails and amphibians, whose last remaining populations on the planet are now being cared for in captivity.[128] Although corals and frogs are worlds apart in many ways, they are alike in how their reproduction is being harnessed to preserve the genetic diversity of their declining populations. Hopefully, aquariums will not become the Noah's Ark of corals, as zoos have become for so many amphibians. But coral scientists may want to start building networks and collaborations with aquarists and hobbyists, just in case this not-so-unlikely scenario indeed transpires.

BEYOND CONSERVATION VERSUS RESTORATION

Right now in the coral reef community there are people who want to do conservation and those who want to do restoration. And I'm saying: you don't need to fight; you have to do both.

—Sarah Frias-Torres, marine biologist, 2016[129]

Coral restoration is a response by a growing number of coral scientists to the impending catastrophe of coral decline due to pollution, overfishing, disease, sedimentation, warming oceans, and acidification. This chapter has documented the history, principles, and debates related to this novel field. Although many of the scientists admit that "coral restoration by itself is not going to change the curve of coral reef decline," in the words of NOAA's Tom Moore, they nonetheless hold that "restoration gives us a fighting chance if and when we fix those global issues."[130]

Yet despite the on-the-ground sense of achievement and the advances in coral restoration, many coral scientists are still reluctant to accept restoration as part of the coral conservation project. For example, Ove Hoegh-Guldberg

explained in our interview that he sees restoration as an outlier to conservation, rather than as one of many tools in the conservation kit. In his words,
restoration is "beyond conservation as we used to know it." From his perspective, then, restoration is a symptom of the Anthropocene, rather than a solution to it. He argues, accordingly, that "we're wasting a lot of money doing this
sort of [management]," clarifying that "I'm not saying that we shouldn't be
trying and refining the techniques. But until we deal with the climate issue,
this is futile."[131] A proponent of restoration, Baruch Rinkevich vehemently
disagrees with these statements. Referring to Hoegh-Guldberg and other
opponents of restoration as "anti-restoration priests," Rinkevich insists that
they are missing the big picture. "I can't say that we are at the point where we
can deliver the entire methodology," he admitted in our interview. "However,
we are responding, step by step, and solving practical and theoretical obstacles
on this long journey."[132]

Sarah Frias-Torres attributes the tensions between conservation and restoration to a generational gap within the scientific community. While the
old-timers "perceive restoration as something bad," she told me, most of the
people working on restoration "are not the old-timers; [they] are people like
me, people who first experienced a coral reef [when] it was already degraded."
She concludes: "I don't know what a healthy coral reef looks like. I've never
seen one. I mean, I've seen pictures, but I've never experienced it firsthand.
Isn't that sad?"[133]

Ruth Gates, director of the Hawai'i Institute of Marine Biology, provides
a similar explanation. But in addition to suggesting that the tensions between
restoration and conservation are the result of generational divides, she also
sees them as deriving from binary worldviews about nature, which she refers
to as backward versus forward looking. On one hand, she told me, "are those
people who are trying to retrieve something and need to go back to this place
where everything looks pristine again"; on the other hand are those who are
saying, "We're here now and we're moving forward." For Gates, the writing is
on the wall. "We need to stop talking about pristine and wild places that
haven't been touched by Man," she said emphatically. "We're already on a fully
engineered planet."[134]

For his part, however, Hoegh-Guldberg refuses to be characterized as a
romantic who is stuck in the past. Coral managers preoccupied with gardening are those who are stuck, he insisted in our follow-up interview. Such coral
management is, in his view, the very opposite of preemptive action; it is an

attempt to depoliticize the contemporary crisis by masking it with temporary fixes and by practicing psychological avoidance and denial. The only reliable way to deal with coral catastrophe head-on, Hoegh-Guldberg has been arguing since the 1990s, is to regulate what is perceived as unregulatable and to control what is seemingly beyond control, or at least beyond the control of scientists: climate change.

The conviction expressed by both sides is stunning, and the arguments compelling. Far from being merely a personal or academic dispute, however, this debate also has significant implications for coral management practices. What is at stake can be gleaned from the following statement by geneticist Mary Hagedorn of the Smithsonian Institution. As the only scientist in the world who freezes coral gametes in gene banks (through a process called *cryopreservation*), Hagedorn was extremely frustrated by what she characterized as the stagnation within the coral community caused by the split between restoration and conservation. In her words,

> Half of the GBR [Great Barrier Reef] has been lost since 1985. The folks who managed the GBR had their head in the sand: until six months ago, they wouldn't even use the word *restoration* in conjunction with the GBR. Very bizarre. When you have such a valuable resource, why wouldn't you want to do everything in your power to maintain that resource? They've never given us any money to cryopreserve the coral on the GBR, either.[135]

The Australian government was not alone in shunning restoration and cryopreservation, Hagedorn complained in our interview. In fact, many conservation organizations "are almost ten or fifteen years behind what's really cutting-edge." According to Hagedorn, this is quite unfortunate. "Today's conservation must be broad, inclusive, and integrative," she stressed.

My first conversation with Hagedorn happened in September 2015. When we spoke again in August 2016, she informed me that she had recently been invited to facilitate training sessions in Australia.[136] In the aftermath of the third global bleaching event, the Australians are now prepared to learn how to cryopreserve their corals, she explained. The urgent plight of corals may finally be bringing together even the most ideologically opposed elements within this scientific community.

Building Bridges and Trees

An Interview with Ken Nedimyer

Ken Nedimyer is founder and president of the Coral Restoration Foundation. He has lived and worked in the Florida Keys for over forty years and has witnessed firsthand the degradation of the Florida Reef Tract. He established one of the largest coral nurseries in the world and has been training restoration groups, especially in the Caribbean, on how to use his unique coral tree technique. Nedimyer won multiple awards, including a CNN Hero in 2012 and a Disney Conservation Hero in 2014. I first interviewed him over the phone on January 4, 2016, then met him in person in Hawai'i, and finally interviewed him a couple of weeks after Hurricane Irma hit the Florida Keys. Nedimyer is the only nonscientist among the interchapter interviews. His narrative is important, in my view, precisely because he is an outsider to that world, therefore providing a fresh reflection on both scientists and the existing legal regimes.

IB: What is the history of your involvement with coral nurseries? How did you become Mr. Nursery?

KN: I was a tropical fish collector in the Keys and did that for twenty-something years. During that period, the reef went from pretty good to pretty bad. It didn't seem like anyone was doing anything about it: the scientists just studied, wrote a paper, then studied something else [and] wrote another paper. Nobody was doing anything with the results. That didn't seem right to me. I was frustrated. I kind of stumbled upon the whole coral nursery thing by accident. We had a live rock[1] farm, the first live rock aquaculture farm on the East Coast. Basically, we got permission to take live stone boulders out and put them on the bottom of the sea and let stuff grow on them for a couple of years and then sell them to the aquarium trade. That was my first foray into open-ocean aquaculture. Nobody had really ever done that before. This was 1994.

IB: Nobody in Florida or nobody anywhere?

KN: As far as growing live rock in the east coast of the Atlantic, what we were doing, nobody had ever done it. One of the things we worked [on] with the state and federal governments [was] to put into the aquaculture program an exemption for any corals that settled on our live rock. . . . [This] allowed us to own any coral that settled on the rock and sell it. One year, we had staghorn corals settle on it. This had only happened one year, one time, for three different colonies. I watched the corals grow and thought, "Oh my gosh, this is what the aquarium grows and sells all the time. I wonder if I could do that in the ocean." We set those rocks with the staghorn corals on them aside and pondered what to do with them. Within a year or so, the corals were way too big to ship, so we just left them there. Then we had storms come through and break them up and I started thinking, "I wonder if we could propagate these things and put these fragments out on other rocks." One thing led to another, and my daughter Kelly and I got this idea of making an offshore coral nursery. We were doing everything underwater: cutting the corals underwater, gluing them onto little rocks. We were going to sell them to the aquarium trade.

We didn't really know what we were doing—we were kind of stumbling forward, watching what the aquarium community was doing. Basically, there are two aquarium communities: the home aquariums (which are the hobbyists), and the public aquariums. I hate to say it, but the hobbyists are way ahead of the public aquariums in what they do. These guys were doing it in their home aquariums all over the place—growing and fragging their own corals. I thought [that] if they can do it in an aquarium, we ought to be able to do it in the ocean. We started and it was very crude, but it was successful. Every time we've done it, we've gotten a little bit better and a little bit more efficient. The process we have now is extremely efficient and productive. It just works.

IB: How easy is the fragmentation process?

KN: It's real easy. They all respond to being cut, or fragged, even brain coral. You can cut a brain coral up into a hundred little pieces with a tile saw and 99 percent of them will live and grow. Aquarists do that all the time. Eventually, we took some of our baby corals into the NOAA Florida Keys National Marine Sanctuary [hereafter "the Sanctuary"] office and asked if we could take them out and plant them on the reef. And they said, "No, you can't do that. There's no permit for that; nobody's ever done that before." We were growing them in the ocean right next to the reef, I explained. Couldn't we just take them from here and put them on the reef? And they said, "Nobody's ever come to us with that question before." So there was no permit available, you couldn't even apply for one. It was all brand-new territory.

Our nursery program kept growing. We had about two thousand corals growing in our nursery, and I was starting to panic because there was no way I could get rid of those corals. I went to the state, I went to the Sanctuary, I went to the Nature Conservancy, Audubon—anybody who'd listen. I'd say, "I've got these corals growing, I've got a whole bunch of them. Anybody interested in doing something with them?" I tried to attract some attention. [Finally, I] got the attention of the Nature Conservancy and they helped get some funding to do some work. Then they got the University of Miami involved, [as well as] Nova Southeastern University, the State of Florida, Mote Marine Lab, and several other people.

So everyone set up their own little coral nurseries that looked just like my coral nursery. It started going from just me, to more than me. They were all doing what I was doing, and meanwhile I was coming up with new ideas. Well, we were getting to the point, by 2005, that we were probably going to run into trouble—and then we got nailed by really bad hurricanes. Our nursery went from two thousand corals to three hundred, so we kind of backed off from doing anything. By the end of 2006 we had a lot of corals again and we were starting [to think] how to get rid of them. And we went to the Sanctuary and basically got our first research permit to put eighteen corals on the reef—which, you know, big whoop-dee-doo—but it was a start. And we put them out on the reef. By the end of 2007 we were getting permits to do another 128.

IB: So tell me, why did you do that?

KN: Why did I do what? Put them on the reef?

IB: Yeah, why did you even bother going to NOAA and persisting when they weren't sure they wanted to do it?

KN: I guess I was clearly seeing that we could do this. Even though we were only [several] years into it, we were having a lot of success, the corals were growing. Corals are these sacred objects in the Keys, like the bald eagle—you just don't go touching coral or doing anything to coral without getting in trouble. You could get arrested for doing what we were doing. . . .

When the Nature Conservancy got involved, we received permission from the Sanctuary to collect new genetic strains, so we added twenty genetic strains to our original three strains. This was in 2005–2006 and was part of this reef resilience program that the Nature Conservancy was working on. The challenge was to come up with a way to tag and keep track of them. We've done a lot of different things. All the corals were mounted on little disks at the time and each disk had a label put on it so when you looked at the coral, you could clean the disk off and you'd find the label and identify which genotype it was. At the time, we were growing the corals on concrete blocks, and all the corals on one block, or on one whole line of blocks, were one genotype. So we kept track of the genotypes that way. And that worked really well until you had a storm, and pieces would break off and you had no idea where they came from. So we created a whole section of unknown corals. Over the years, we accumulated a lot more of them. Eventually, we just [outplanted] a reef of one thousand unknown corals. We have no idea which ones they are, but we have a reef.

As for the permits—we went from getting eighteen to 128, then to five hundred and to twelve hundred. When we went from twelve hundred to fifty thousand—that's when they choked. At that point, it was clearly not a research project anymore. It was full-on restoration. It took us a year and a half to get that permit. That was a restoration permit. The problem was that nobody had ever done that before: there were no standards, no protocols, no best practices, nothing. It just hadn't been done, at least in Florida or in the National Marine Sanctuary Program. So we were working with the Nature Conservancy, various programs within NOAA, and the State of Florida to try to develop best practices for nurseries and best practices for restoration. They were really reluctant to give out these restoration permits.

IB: How do you explain their reluctance?

KN: In the past, all kinds of well-meaning restoration plans have gone bad. The classic example is the salmon on the Pacific coast of the United States. They were going to try to restore salmon populations, so they started collecting all these salmon from these different streams and culturing them in one location and then trying to release them. Little did they know that these salmon are hardwired to go back to their native stream: if the salmon spawns in stream A, it will go out into the wild and swim around for a few years and finally go right back to stream A and lay eggs again. So the whole system on the West Coast of the United States got screwed up by well-meaning restoration work. They did what seemed like the right thing to do, but at the end it was a disaster. They just didn't want to see that happen with corals in the Keys. So we had to demonstrate that what we were doing was not salmon and that it was not going to harm anything. We had to have a plan, and it was a lot of work to get them to buy into it and accept what we were planning.

When you come to a river or stream you have to build a bridge to go over it. But once you've built the bridge everyone can cross. Whether it's building a nursery and fragging corals, or collecting them and demonstrating that you can do it, or planting them on the reef—all of these different things, once you've done it and created the permitting path for it, then it's easy to get the same permit again. And we've done a lot of that, and are still doing it. Once we outplanted the fifty thousand corals, we got an amendment for 120,000, and we're just about done planting those. So we have a ton of coral growing out there. We have a permit that's been in the system now for nine months and we're going to outplant 350,000 corals. I just don't think [NOAA] realizes the scale we've reached with this project.[2]

IB: How many nurseries and corals do you have at the moment [2016]?

KN: We have five nurseries in the Keys right now. We're going to be putting in another one. Our biggest one is still the original one and there are almost five hundred coral trees there (figure 26). Each tree can hold

FIGURE 26. Ken Nedimyer tends to *Acropora cervicornis* (staghorn) corals at the Tavernier Nursery in Florida, April 2011. Photo © Stephen Frink.

one hundred corals. The other four nurseries have roughly fifty trees in them, so about five thousand corals each. The trees each have ten arms and each arm has holes, ten holes make for ten corals in the arm. The smallest coral we might hang is fifteen or twenty centimeters long—that's a staghorn coral—and when they're about a hundred centimeters long, we cut them off and plant them on the reef. So at any given time, one-third of our corals might be twenty centimeters or less, one-third are twenty to a hundred centimeters, and then the last third are huge—a hundred centimeters or more.

We've also added elkhorn corals. The staghorn was doing great but I really thought elkhorn was the most important species to be working with—that's the coral that builds the main reef. If you look at the barrier reefs and the fringing reefs anywhere in the Caribbean—they were all built by elkhorn coral. It's a hardy, fast-growing coral that builds massive reefs. And it is in worse shape than the staghorn coral. I applied for several different grants to fund an elkhorn nursery [but] I couldn't get any traction, nobody wanted to fund it. "That won't work, you can't do that," I heard again and again. But I knew we could. In 2009, I got permission to do it anyway, I just paid for it out of pocket. I thought, "Look, I'm going to show you how to do this, I'm going to do it." So we started an elkhorn nursery in 2009 and put a lot of effort in that over the last several years. Now, one hundred out of five hundred trees at our big nursery are elkhorn trees, and the other four hundred are staghorn. One nursery is just elkhorn, and the other three nurseries are about half staghorn, half elkhorn. We have close to sixty different genotypes of elkhorn in our nurseries now.

We were also going to collect some of the [coral] hybrids in the Keys. There aren't very many of those here, so we were going to collect and culture them. But [NOAA] just shot it down. They said they wanted to maintain pure strains, and didn't want to mix them. I'll probably do it on a research basis with a research permit, but they're not going to let me transplant the hybrids onto the reef, mostly because of the Endangered Species Act. It's a long story. But right now, the hybrids are being pulled out from our permit. That kind of caught me off guard.

IB: It sounds like you just got yourself another bridge to build . . .. Taking one step back, though, how did you come up with the unique idea of designing tree structures for propagating corals?

KN: Our idea of a coral tree evolved from the line nursery. There were people already using line nurseries in the Caribbean. It looks like an old-fashioned clothes line. You have an anchor at the bottom and a line with a heavy, big float on it, then twenty feet away you have another one and between those two vertical lines you tie horizontal lines and you hang the corals from the horizontal lines. We started some line nurseries in our nursery and we thought, "Wow, this is great." The corals grow two or three times faster than they do on the ground, they are susceptible to fewer diseases, it's more storm proof because they sway in the storm instead of trying to resist it, and there is a whole lot less maintenance. So we went to the Sanctuary and asked to put twenty of them here and twenty of them there and twenty of them here. And all of a sudden somebody in the U.S. Fish and Wildlife Service who was reviewing the application decided that these line nurseries pose an entanglement hazard to protected marine mammals and turtles. So the Sanctuary said no, we're not going to let you do that. . . .

The day we got that decision from the Sanctuary was a cold and windy February day. We weren't going to be diving. We'd had a couple cups of coffee and we were agonizing over this decision that had just been handed down. We started dreaming up: how could we still suspend the corals without posing an entanglement hazard? We just started tinkering around with PVC pipes and talking and we came up with this idea of a tree with these arms with things dangling from the arms that don't really pose an entanglement hazard. We've gone through seven or eight modifications to the tree concept since we started, but it is definitely the way to go. Every tree can have its own genotype, or you can put multiple ones on a tree [and] just label the branches. An easy way to organize them, an easy way to keep an inventory, and a great way to lay out a nursery. I've looked at what a lot of other people are doing and some people are doing things that work great in an area that has a lot of protected water or some place that's real shallow, where you can't really design a tree like that. But for what we're doing in the Keys and what we've done in the Caribbean, it seems like the tree is the best solution. A lot of people have copied it and it seems to be successful.

When we came up with this tree idea, people told me that I need to consider [patenting this invention]. But I said, "You know what, I think this is a breakthrough technology. I think this is going to revolutionize our ability to grow corals. And I want anybody and everybody to have it if they want it." So we published a paper and just basically gave it away. We said "Look!

Here it is, everybody! Do it, modify it, do whatever you want. You don't have to ask for my permission, but here's something that just might change the way you grow corals." So that was the way I approached it.

There are other designs out there. Table nurseries have a frame made out of a rebar rod that's on the bottom, then between the frame are sometimes lines or ropes, and the corals are hung between the ropes or even inside the rope itself. Then there's another one called a midwater nursery, where they have big nets suspended in the water and the corals are grown on little plates or trays on top of the nets. They've done that in the Red Sea. You could never do that here, there's too much wave energy.

IB: So how many nurseries are there in Florida?

KN: There are currently about ten or twelve nurseries in the Florida Keys. We have five, Mote has one, the Nature Conservancy has one, Florida Fish and Wildlife has two, University of Miami has two, NOVA Southeastern has one, and there are a couple other tiny nurseries. Our large nursery alone is larger than everybody else's combined. Outside of the coral farms in Indonesia, it is the largest nursery in the world.

IB: What is the difference between a farm and a nursery?

KN: I believe they are two completely different things. In my view, a nursery is for restocking and replanting, a farm is for growing something to sell. As it turns out, we never sold any corals. I joke about it because I finally said to my daughter: "Kelly, we're just going to give these all away." And she was like, "What? Are you crazy?" And I said, "Yeah, I am." I used to tell her about all the money we were going to make, because we'd be the only people in the world with that particular species. But I walked away from making money. We never sold the first coral and we gave up that right. When the Protected Resources Division of the National Marine Fisheries Service was writing up the rules for staghorn and elkhorn corals under the Endangered Species Act, they actually came to me and said, "Ken, you technically own those three genotypes and, if you want, we could make an exception that you could still sell them." And I thought about it and said, "No, I don't want to exercise that exception." I just thought it would be a lot cleaner if we didn't have that, so there's never any question about what I was doing with any of the corals. Sometimes I kick myself for that decision, but I know it was the right thing to do. So we never sold any coral, and Kelly went on to become a dentist.

IB: Do you know how many corals you've restored over the years?

KN: I replanted a lot of corals onto twenty-five or thirty reefs. I can't say that these reefs are restored to what they used to be, no. But they're in the process (figure 27). I really am more excited to work with elkhorn than with staghorn. While the staghorns come and go pretty quickly, the elkhorn is a long-lived, enduring coral. But it's still early for us, we haven't outplanted that many elkhorn corals. They're hard to transfer, they're hard to keep alive—they're very difficult.

IB: Have you considered protecting the coral genotypes propagated in your nurseries?

KN: We've been working with the [Florida Aquarium]. They're going to be like a gene bank for us and will have a copy of everything we have in our nursery in their greenhouse. Because if we got hit by a category 4 or 5 hurricane, we might not have much left.[3] So it would be really nice to have a backup set of corals somewhere. We're also going to have replicates of the genotypes from each of the Keys' nurseries among the other nurseries in the Keys. But I would still feel more comfortable if we had some in Atlanta and Tampa.

We're also working with Mary Hagedorn on cryopreservation. We froze sperm last summer and thawed it out to fertilize eggs and it worked, so it's kind of exciting. We now have about 150 genotypes of staghorn in our nursery. And we're doing sexual reproduction, too. When it really comes down to it, sexual reproduction is a whole lot easier than growing corals asexually. And the returns are better for the long term, because we can cross a heat-resistant coral with a disease-resistant coral and may get something that has both properties. That's our hope. At this point, we're good at getting them to fertilize, we're good at getting them to settle, but we're not having a lot of luck getting juvenile corals all the way through the process. But we're getting there.

IB: What are some of the challenges with sexual reproduction?

KN: The hard part is that with these corals, you only get one shot at it per year, so it takes many years to develop the techniques. We're spawning them in the nursery, with nursery-raised coral. The Florida Aquarium [people] come down every summer—this last summer was probably the sixth summer they've come down—and we select the different genotypes. We test them underwater, we collect all the eggs and sperm, we bring them all back to shore and mix and match and do whatever else and then try to raise them on shore, primarily at the Florida Aquarium. But they just haven't had the best of luck.[4] We've gotten really good at predicting when they're going to spawn and catching them and harvesting all these different genotypes to try different things with them. People could never do it before because you almost couldn't find enough staghorn coral spawning anywhere to do it.

IB: Have you been invited to spread the restoration word elsewhere?

KN: Well, we train people to do restoration in other Caribbean locations. We provide the material and teach them how to do everything. We've

FIGURE 27. Coral Restoration Foundation intern James Boyle monitors a group of previously outplanted *Acropora cervicornis* (staghorn) corals, also called a "cluster," on Alligator Reef, Islamorada, in Florida, 2016. Photo by Jessica Levy, Coral Restoration Foundation.

helped locals set up nonprofits in some of these different islands. First, we go in and show them how to do it, where to find the corals, how to handle them, where to set up the nurseries, how to take care of them, how to plant the corals, and all that. Then we work with them on how to run a nonprofit to support this work. We've gone to different places like Bonaire, Curaçao, Jamaica, and Roatán. We've also been approached by people in the Philippines and Indonesia—but we're just not quite ready to go there.[5] The Caribbean is in a lot of trouble, more trouble than anywhere else. And I know the Caribbean—I know the coral, I know the history, I know the issues better. So for now, it's easier for us to support work in the Caribbean. We're really trying to develop this model in the Caribbean before we go elsewhere. It's tempting, people are begging me to go to Kenya and other places. And I want to go there, I want to do it, but I also know my limitations.

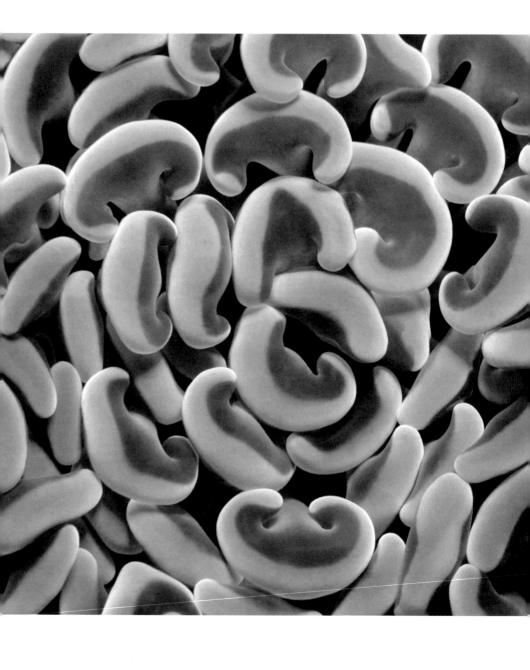

CHAPTER 4

Coral Law under Threat

Climate change challenges the capacity of law.

—R. Henry Weaver and Douglas A. Kysar[1]

Law is constantly colonizing catastrophe, reframing it as injustice, expanding the bounds and jurisdiction of law, and consequently expanding the zone of human control and responsibility.

—Linda Ross Meyer[2]

BREATHING LIFE INTO LAW, AND LAW INTO LIFE

For me, diving is a deeply meditative experience. The background noise of terrestrial life is muted and one automatically attunes to the breath. As in traditional meditation practices, underwater the importance of the breath is acknowledged more directly—one can't help but realize how indispensable it is for life. Breathing is minimal and crucial; it is the source of motion. Inhaling takes you up, exhaling brings you down: in-up; out-down. Time is defined by the movement of the breath. The present moment unravels, one breath at a time. "Of the candidates for life's essence, breath is strongest," wrote evolutionists Lynn Margulis and Dorion Sagan. "So long as there is life, there is breath. Breath is invisible. Like wind, it moves things."[3]

Seeking the immensely colorful corals from the Red Sea expeditions of my youth, in 2015 I traveled to the Great Barrier Reef in Australia, one of the world's acclaimed wonders—a designation that carries with it its own legal administration. On the long ride from Port Douglas to the outer Agincourt

---

FIGURE 28. *Euphyllia ancora*, or hammer coral. Courtesy of Coral Morphologic.

reefs on the fancy tour boat, *Quicksilver*, I had plenty of time to chat with one of the instructors about the state of coral reefs around the world. He has seen it all, and, at least from his perspective, the future of tropical corals was looking quite grim. At the same time, he was excited to tell me about his involvement in coral conservation. It was from him that I first learned about the "pest" crown-of-thorns starfish that has been devastating Australia's reefs, and about the national project for the eradication of this species. He had personally killed dozens of these starfish by lethal injection, the instructor boasted without missing a beat. From the point of view of Australian environmental law, this national eradication project is not only permissible but also desirable, thereby revealing the underlying order that organizes the value of life and death in relation to corals, a project concerned more generally with "making live" and referred to by Michel Foucault as *biopolitics*.[4]

I finally returned to the Red Sea in June 2015, expecting the worst after twenty-five years away. Although not too long ago this would have been but a flicker of time in the life span of corals, who can live for hundreds if not thousands of years, time has come to matter quite differently nowadays. Before I arrived, Assaf Zvuloni, my host and the manager of Eilat's Coral Beach Nature Reserve, informed me that the Japanese Gardens, the site of my first scuba dive experience, had been closed to the general public. The corals were not doing so well, he explained, so the park's manager at the time decided to declare this a "no go" zone, which meant that any direct uses, including recreational diving, were now prohibited. Instead, Zvuloni took me and my older daughter, along with his own children and friends, to snorkel over a nearby reef. His wife, also a marine biologist, had a high fever and a serious cough that afternoon, but insisted on diving with us. "The ocean is my cure," she cut me off when I tried to convince her to stay on shore. It was a beautiful dive, replete with an octopus spraying ink at us and a display of territorial aggression by two male parrotfish. Still, I couldn't help but feel sad. When my daughters reach my age, will there be any corals left for the law to protect with "no go" zones? And how effective are these local prohibitions when other, much larger, problems loom over the continued life of reef-building corals on this planet?

Although many of the scientists I interviewed found it hard to believe, much of my academic training occurred in the classrooms and corridors of law schools. Upon completing my legal training in Jerusalem, I worked as a state prosecutor, then as a lawyer for an environmental nonprofit, and, after a few detours, eventually focused my master's and doctoral dissertations on issues of land and law. I'm particularly fascinated by how law protects and promotes life—law's "making live"—and, in this case, by how law regulates the life of certain corals. I am equally interested in how, at the margins of such life-making practices, law also decrees, and justifies, death. The government-orchestrated killing of the crown-of-thorns starfish by both humans and robots will serve as a case study for this form of "making die." Although many of the scientists I interviewed for this project insisted that their work has little to do with the law, a mere scratching of the surface has revealed law's prominence in their work—if not formally, then informally through guidelines, procedures, and standards for management and protection.[5] Science and law are entangled in many ways, as this chapter will show.

Specifically, the chapter will discuss the myriad legal regimes that pertain to reef-building corals and, mainly, those that govern national waters. I will start by asserting that law is neither a blank nor a neutral slate. In the context of oceans in particular, Western law has long played a central role in colonial expansion and in the exploitation of marine resources. Under the guise of pleasant-sounding doctrines such as the Freedom of the Seas—a principle put forth in the seventeenth century, essentially limiting national rights and jurisdiction over the oceans to a narrow belt of sea surrounding a nation's coastline, with the remainder of the seas (the "high seas") proclaimed to be free to all and belonging to none[6]—imperial powers generated a multitude of claims, counterclaims, and sovereignty disputes. The role of law in the ongoing imperial and capitalist exploitation of marine resources is especially relevant at this time, when the Freedom of the Seas, as well as other legal doctrines, are being reinterpreted to enable the mining of the seabed, offshore drilling, and deep-sea fishing.[7]

The imperial biases of Western ocean law were especially emphasized by ecologist Peter Sale, who insisted in our communications that environmental law is profoundly un-environmental. "Time and again, 'environmental' law is found to favor individuals and corporations doing business that has degrading environmental byproducts, rather than favoring the environment that the law is theoretically there to protect," he told me.[8] While "the cocktail of human stresses is mixed differently from place to place," oceanographer Callum Roberts writes more broadly, "the results are the same. . . . We are transforming life in the sea, and with it undermining our own existence."[9] We have forgotten that even our own breathing begins in water. Our laws, the formal ones at least, often reflect this collective amnesia all too strongly.[10]

The historical analysis of the nexus of oceans and law and their contemporary manifestations in the high seas are of critical importance, but they are not my main focus here.[11] Instead, I am concerned with the more mundane and contemporary entanglements between coral life and law—law's breathing life into corals, if you will, by protecting their continued existence, and coral's breathing life into law by providing it with the materiality required for governance. Indeed, corals are saturated with law, which protects, ignores, or damages them—or any combination thereof; they are also saturated with science, which names, classifies, and lists them, thereby making them legible, or illegible, to the law. In this sense, corals are life-law-science *coralations*, a term that encompasses their "coproduction"[12] as well as other, perhaps less mutualistic, relations—including tensions and even negations.

Immersing myself in coral environments has taught me that law is also capable of breathing organisms *out* of existence. This is the case with the corals' "enemies," the crown-of-thorns starfish and the lionfish, and even with certain kinds of corals, such as hybrid and introduced corals as well as deep-sea corals, who have each been made invisible to, or an exception to, the law. This chapter will draw on multiple interviews with administrators and managers to highlight, in particular, how the traditional focus on species by conservation laws such as the U.S. Endangered Species Act disadvantages corals and, relatedly, how ill-equipped such laws are when dealing with the imminent threats to coral life on a global scale. While this fundamental incapacity of law to see, manage, and protect coral is, for the most part, a systemic failure that has resulted in much frustration and despair toward the law on the part of coral scientists, law can also be used as a tool for resistance and change, thus providing hope for corals in the Anthropocene.

ALGAE AND CORAL LAWS: A GENERAL INTRODUCTION

Mary Alice Coffroth and I teach at the same university, which is hours away from any ocean. It was from her that I first heard about the wonders of coral algae. Apparently, before the advent of genetic sequencing, not much was known about these organisms, who were historically classified in the plant kingdom and later shifted into the kingdom Protista, a catchall for a diverse group of loosely related and typically microscopic organisms. More recently, geneticists have been classifying algae into clades and dividing them into A, B, C, D (up to H) branches, which, as one coral taxonomist told me, are the rough equivalents of genera.

Coffroth had plenty of stories to tell me about how law matters for corals, and one in particular made her choke with laughter. She recounted that in the 1970s, when all legal permits were handwritten, she returned with a group of scientists from a research trip in Mexico, their trunk stacked to the top with coral algae. Unfortunately for them, they had one permit and two cars, which would have been fine except that they had lost each other during the trip. When the permit-less car reached the border, the officer took one look at the trunk and asked to see their federal permits for transporting plants. "These are not plants," the scientists assured the officer. "They're algae." Puzzled, he let them pass, an incident that Coffroth found hilarious.[13]

Working with biologists for nearly a decade, I long ago resigned myself to not understanding their jokes. I end up looking up too many terms and, by the time I catch up, the party is over. Coffroth explained during a later interview that it's still amazingly simple to transport symbiotic algae across state lines. By contrast, one cannot transport the coral animal, dead or alive, without adhering to strict requirements and obtaining special permits.[14] Whether an organism is an animal, a plant, or neither thus determines whether and how far it can be legally transported, by whom, and for what purposes. And while myriad coral species are listed as endangered and protected, algae are not—despite their importance for healthy coral reef systems. These are just some of the ways in which laws prioritize higher forms of life, priorities that don't merely stay on the books but also play out in practice. One admittedly marginal manifestation of this practice was that I have been able to observe, under the microscope in Buffalo, New York, symbiotic coral algae from the Atlantic Ocean.

In light of the poor and declining state of coral reefs around the world, and especially in the Caribbean, where many coral reefs under U.S. jurisdiction are located,[15] I didn't expect to find many laws that set out to protect them. I was wrong: a vast array of formal laws, orders, and agencies govern corals in the United States alone. These legal regimes can be divided into two separate areas of focus: direct protection and pollution prevention.[16] The establishment of marine protected areas exemplifies the traditional form of protection for coral ecosystems. However, nearly all of the marine protected areas in the United States allow multiple uses, including fishing. Some areas, such as marine reserves, are more restrictive. Marine reserves with "no take" policies occupy about 3 percent, or four hundred thousand square kilometers, of U.S. coastal waters.[17] This means that the other 97 percent of marine ecosystems, which also frequently contain valuable and endangered lives and habitats, are not provided the same level of protection. Finally, while legal protections exist on the books, they have many loopholes and exceptions that render them much less protective in practice.

The disparity between "law in the books" and "law in action"—a distinction coined by Roscoe Pound in 1910[18]—is relevant also in the case of the Magnuson-Stevens Act of 1976,[19] which provides a broad platform for the federal regulation of marine fisheries in U.S. waters, including corals. In 1998, President Clinton established the Coral Reef Task Force to protect and conserve coral reefs.[20] The Coral Reef Task Force was deemed responsible "for mapping and monitoring U.S. coral reefs; researching the causes of coral reef degradation including pollution and overfishing and finding solutions to these problems; and promoting conservation and the sustainable use of coral reefs."[21] As a principal member of the Coral Reef Task Force, and as directed by the Coral Reef Conservation Act of 2000,[22] the National Oceanic and Atmospheric Administration (NOAA), an agency within the Department of Commerce, was tasked with spearheading the efforts to conserve and protect coral reef ecosystems.

In 2000, the Coral Reef Conservation Act established the Coral Reef Conservation Program, a body aimed at coordinating internal efforts within or by NOAA "to save coral reef ecosystems by focusing on climate change, land-based sources of pollution, and unsustainable fishing practices."[23] Other specific programs established within NOAA include the National Marine Fisheries Services' *Acropora* Recovery Team, which promotes conservation of elkhorn and staghorn corals through a recovery plan designed by multiple experts;[24] the Coral Genomics Initiative, which aims to "study the genomic mechanisms of coral resilience against future thermal stress and ocean acidification";[25] and the Coral Disease and Health Consortium,[26] a voluntary and interdisciplinary working group within the Coral Reef Task Force that focuses on understanding coral health and disease. Finally, a network of field-based coral nurseries, operating through NOAA's National Marine Fisheries Service (NMFS) Restoration Center, focuses on restoring four main fish habitats, including corals.[27]

Despite this mushrooming of administrative bodies and initiatives, the NOAA administrators I spoke with have often complained about lack of manpower, funding, and authority. As a result, the work of identifying imperiled corals—and, for the most part, also that of restoring them—has increasingly fallen on nonprofits, individuals, and even corporations, with NOAA hobbled by outdated laws and practices and racing to catch up (as discussed in the interview with Ken Nedimyer that preceded this chapter).[28] Finally, despite their rapidly deteriorating state, help for imperiled coral species can be tied up with red tape and coated in dense legalese for many valuable years.

Nonetheless, every now and again there are signs of progress. For example, in 2016 the Coral Reef Task Force released the *Handbook on Coral Reef Impacts: Avoidance, Minimization, Compensatory Mitigation, and Restoration.* This handbook is a first-of-its-kind review of federal, state, and territorial authorities and responsibilities as well as a compendium of current best practices and science-based methodologies for quantifying, assessing, and mitigating impacts to coral reef ecosystems and restoring these systems, including the use of appropriate compensatory action to replace lost coral "functions" and "services."[29] The report, which signals a shift in legal culture from preservation toward compensation and restoration, admits that, to date, there is no universally accepted model for determining the level of compensation for inadvertent damage to coral reefs and points to the challenges of applying existing compensatory models, such as those developed for wetlands, to coral ecosystems.[30] Recent years have seen the growing utilization of such economic tools for compensation and mitigation and their intensifying deregulation and reregulation under the law.[31]

LISTING AND RECOVERING *ACROPORA* CORALS

Typically viewed as the strongest federal environmental law in the United States and possibly in the world, the Endangered Species Act of 1973[32] is aimed at preventing the extinction of imperiled plant and animal species. To date, twenty-two coral species within U.S. jurisdiction have been listed as threatened and three species outside U.S. jurisdiction have been listed as endangered under this law. Such listings carry profound material effects: once a species is listed as threatened or endangered, one cannot touch, let alone move or break, any of its members without first obtaining a permit from either the U.S. Fish and Wildlife Service (for terrestrial species) or NOAA (for marine species). Other federal government agencies are required to consult with NOAA if their projects might impact any listed marine species, and the habitats of these species may be designated as critical, which entails special management and protection. NOAA is also mandated to design a recovery plan for each listed species.

The first two coral species listed under the Endangered Species Act were the elkhorn and staghorn corals—*Acropora palmata* and *Acropora cervicornis.*

The genus *Acropora* is the most abundant and species-rich group of corals in the world. Elkhorn and staghorn corals are two of three acroporids found in the Caribbean Sea and the Atlantic Ocean (the third being the hybrid *Acropora prolifera*, discussed below) and were historically among the most dominant reef-building species in the region. Declines in their abundance have been estimated at 97 percent. On May 9, 2006, the NMFS—also known as NOAA Fisheries—published its Final Rule listing the elkhorn and staghorn as threatened under the Endangered Species Act.[33] It also assembled the *Acropora* Recovery Team and developed a recovery plan for these two species, including the designation and protection of three hundred square miles of critical habitat in U.S. waters in 2008.

Recovery plans, as defined by the Endangered Species Act, establish the goals that must be reached before a species can be removed from the list and outline concrete actions that are perceived as necessary to achieve these goals. While the legal reach of the Endangered Species Act is very broad within areas under U.S. jurisdiction, outside of this jurisdiction it applies only to U.S. federal agencies that are funding or carrying out actions that impact listed species.[34] The idea behind this extra territorial purview is that many threats are regional or even global, and the abatement of these threats within U.S. jurisdiction alone will thus not suffice for their recovery. Consequently, the recovery plan for elkhorn and staghorn corals includes populations of these species outside the United States and recommends cooperation with other countries throughout the Caribbean and Atlantic regions.

Although such broad inclusion may sound sensible on the books, it has led to "absurdity in practice," in the words of Peter Sale.[35] Sale pointed out that this recovery plan, which stretches over 167 pages and took five years to produce, is anything but practical. He was particularly dismayed by its summary, which states that "the total cost of recovery is not determinable given the global scale of many of the threats impeding recovery," while at the same time estimating the cost within U.S. jurisdiction alone at $254 million and adding that "this represents an extreme underestimate for the actual cost of recovery."[36] The plan also estimates that "it will take approximately 400 years to achieve recovery."[37] The impracticality of this recovery "plan" has elicited criticism on the part of many scientists, who have expressed a general frustration with the legal tools available for coral protection. Sale asks accordingly,

Could not all the money spent on this document have been better used to actually manage [the] environment? But that is what seems to happen frequently—bureaucrats follow regulations set out in legal instruments, without bothering to lift their heads and say, "Wait a minute, why are we doing this? Let's have a law that actually makes sense."[38]

Law and coral recovery are also entangled in other, perhaps more practical, ways. Zac Forsman, a geneticist and restoration scientist, was involved in a large coral damage case in Hawai'i where, in 2009, a ship grounding destroyed about twenty acres of reef.[39] Under the U.S. Clean Water Act, if a certain number of "coral-colony-years" are lost, they must be replaced.[40] Coral-colony-years are calculated based on the number of coral from each species and the number of years it would take them to return to their original size. For example, a one-hundred-year-old coral would represent one hundred coral-colony-years. Using a tool called "Habitat Equivalency Analysis," the U.S. Fish and Wildlife Service determines how to replace the lost corals and how to calculate the cost of this process. As Forsman explained in our interview:

> For a year and a half I worked with U.S. Fish and Wildlife Service on proposals for how to replace coral-colony-years. I wrote proposals, including a proposal for the urchin hatchery and the coral nursery, that helped to estimate the cost of replacing the coral-colony-years that were lost. When a navy ship accidentally ran a ship over the reef, those proposals helped to settle the case for about seven million dollars. If the coral nursery that I helped build can use mass production to grow coral seedlings and plant them successfully, it may help recover many of these lost coral-colony-years.[41]

While this instance of law-science collaboration for coral recovery is presented as a success story, it is important to remember that its practical relevance is limited to those instances where the perpetrator is easily discernible, and can be made accountable.

Beyond the practical concerns with this form of coral offsetting, or exchange, one could also argue that it undermines the life of the corals who were killed and minimizes the act of killing them. In other words, by abstracting actual coral lives to "coral-colony-years," regulatory practices render "bits of nature exchangeable with one another."[42] Geographer Aurora Fredriksen calls such offsetting and banking practices "new markets of biodiversity" and claims that they tend to sacrifice the individual for the sake of the species. In her words, "in performatively subverting the value of individual organisms to the species unit, biodiversity conservation practices position individual animals as interchangeable with other individuals who are categorised as members of the same species."[43] Environmental anthropologist Sian Sullivan expresses a similar concern that "diversities are lost in the world-making mission to fashion and fabricate the entire planet as an abstracted plane of (ac)countable, monetizable and potentially substitutable natural capital."[44]

LEGAL STRATEGIES FOR CLIMATE LITIGATION

In 2009, the Center for Biological Diversity—a nonprofit science-law collaboration with a strong, some might say aggressive, litigation focus—petitioned NOAA to list eighty-three coral species under the Endangered Species Act. In 2014, the Final Rule listed twenty coral species[45] and also relisted the two *Acropora* species that were already listed as threatened.[46] Shaye Wolf and Miyoko Satashita, climate science director and oceans director at the Center for Biological Diversity, drafted the petition. In our interview, Satashita summarized her perspective on the process. Species whose habitats are designated as critical are twice as likely to recover as those that aren't, she explained, expressing the widely held belief of scientists in the power of legal norms and arrangements to make a difference for corals and other species (which often results in much frustration, as we have already seen and will see more of shortly). She also mentioned the wider implications of this listing and its resulting recovery plan: "One of the very notable things about this recovery plan is that it set up clear criteria and targets for ocean acidification and for global warming [that are] needed to conserve the corals and recover them."[47]

Many foreign climate advocates and scholars are shocked to discover that the United States—a country with highly developed environmental laws—does not have a federal climate change statute. Instead, government officials and legal advocates are forced to cobble together bits and pieces of other environmental laws to try and address some aspects of U.S. greenhouse gas emissions.[48] With this in mind, the Center for Biological Diversity's underlying agenda is clear: they would use an existing law, focused on species conservation, as a strategy for imposing climate change regulation on the federal government. Legal strategies like this can transform corals from passive victims of climate change to important actors in combating it.

But the heroic stance of fighting climate change through coral recovery criteria did not seem so heroic from the standpoint of the administrators charged with implementing these protections. The arduous work of compiling data and conducting public hearings for the Endangered Species Act's listing petition was performed by NOAA, with only two people on staff for this process. Alison Moulding, a NOAA natural resource specialist, told me in 2015 that after her agency was sued by the Center for Biological Diversity to issue the proposed listing rules for the eighty-three coral species, NOAA officials couldn't work on anything else until the listing was fully researched and finalized, which took years. Unfortunately, she said, this also came at the detriment of coral recovery, which has no strict deadlines and so was easier to postpone. "When something goes into litigation, it's not the lawyers doing all the work," Moulding complained. "It's the biologists."[49]

The disparate perspectives of scientists and lawyers have also resulted in tensions and disagreements about how best to use the law. During the International Coral Reef Symposium in Hawai'i in 2016, the Center for Biological Diversity arranged an invitation-only meeting for biologists and lawyers to jointly brainstorm innovative legal tools for coral conservation. All the relevant NOAA people were there, and so was the point person on corals from the International Union for Conservation of Nature (IUCN), as well as a few others. After numerous Skype interviews, telephone conversations, and e-mail exchanges, this was my opportunity to finally meet these people in the flesh and to see them interact with one another in one shared space. Sitting there, I reflected on the complex relationship between the scientists and the lawyers in the room. Despite the common agenda and the emotional bond toward corals that were genuinely shared by all, these two groups seemed extremely cautious, if not suspicious, of one another. Based on later conversations, I inferred that while the lawyers seemed too belligerent and single-minded to their scientist peers, the scientists seemed too hesitant and risk averse to the lawyers.

TO UPLIST, OR NOT TO UPLIST?

Conservation scientists and environmental lawyers also don't always agree on the efficacy of existing laws, exemplifying that even the seemingly straightforward act of listing a species is not so straightforward after all. In response to the 2009 listing petition by the Center for Biological Diversity, NOAA proposed "uplisting" the elkhorn and staghorn corals—the two species already designated as threatened in 2006—to the status of endangered. This would entail more stringent protections against what the Endangered Species Act refers to as a "take," which includes any handling of, harassment of, or impact on the listed corals. I had expected the various conservation groups to support, and even fight for, such increased legal protection. Instead, they opposed NOAA's proposal, and it eventually fell through. Legally, the elkhorn and staghorn corals remain threatened and are not endangered.

I will suggest here that the uplisting proposal and its failure resulted from fundamental differences in approach between conservation scientists and lawyers, thus highlighting the ongoing tensions between these communities, even as they operate under one agency. Margaret Miller, a research ecologist in NOAA's Protected Resources Division, explained the dynamics. "It takes at least a year to get a research permit for any species listed as endangered," she said. "And that, clearly, would have been a hindrance to a lot of [existing research and restoration efforts]. In that sense, it was beneficial not to [uplist] these corals."[50] Nurseries and other restoration initiatives that have been obtaining take permits from NOAA were especially concerned that their fragmentation and outplanting operations (i.e., transplanting corals from nurseries onto the reefs) would be hindered or at least stalled. From their perspective, stronger legal protections would actually result in lower levels of restoration and could thus end up harming rather than benefiting the corals.

Miller was therefore relieved when her agency decided not to uplist the two coral species. "At least [one] bureaucracy that we don't have to deal with, thankfully," she told me.[51] She was not the only scientist I spoke with who expressed this view. Nilda Jiménez-Marrero, a biologist working for the Puerto Rican Department of Natural Environmental Resources, has been facilitating a network of coral nurseries across Puerto Rico. From her perspective, the proposed change in status for *Acropora* corals would have introduced a huge burden to the Caribbean nurseries. "A nightmare," as she put it.[52] The aquarium industry was also concerned. Andy Rhyne of the New England Aquarium recalled that "the aquariums literally said that they were no longer going to work with this species if NOAA listed it as endangered because it is so laborious to fill out their Section 10 paperwork."[53]

Contrary to the arguments against the proposal, Jennifer Moore, marine conservation program manager at NOAA, believed that the uplisting was important and should have moved forward. "There was a lot of confusion and, in my opinion, misplaced fear, by the scientific community that an endangered listing would prevent the types of research and restoration activity that need to occur," she explained. Yet she assured me that NOAA's legal department was "working on a programmatic permit so that research and nurseries could continue completely unchanged from day one of the potential 'endangered' listing."[54] According to Moore, the number of take permits would actually have increased, rather than decreased, with the uplisting[55] because an endangered status tends to attract more attention and funding. "Nobody believed NOAA's line of argument here," Rhyne responded. "We were worried about the lawyers."[56]

The miscommunication over what could otherwise have been handled as a straightforward question exposes the complicated dynamics between conservation managers, lawyers, aquarists, and federal agencies as they negotiate the application of environmental laws. Unfortunately, these tensions often play out to the detriment of corals, as they sow division rather than unite the experts who care about them.

CLASSIFYING CORAL: HYBRID 83

Despite some stark differences in their approaches, lawyers and scientists also share much in common. Centrally for this book, these two groups tend to obsessively classify their subjects of research and regulation.[57] Moreover, legal regimes often rely heavily on scientific classifications, especially in the field of environmental law. The scientific project of classifying all living things into species, in particular, is essential for the regulatory project of monitoring and protection. However, corals often don't fit properly into existing categories and classifications, giving aquarists and managers a big headache and forcing lawmakers and administrators to rethink their assumptions. Jennifer Moore of NOAA explained the problem: "Really, the more we learn about coral taxonomy, the more [we] realize it's very fluid, very plastic. There aren't the same hard boundaries between species that we see in other organisms like vertebrates." Coral species "split, come back together, split again, [and] come back together over evolutionary time. It's just not as clean [as Darwinian models have it]," she told me.[58]

The story of unlisted hybrid coral number 83 demonstrates the importance of classification generally, and the power of ideals about purity in particular,[59] for the daily operation of environmental laws. Coral 83, a hybrid of the already threatened elkhorn and staghorn corals, is also known by its scientific name *Acropora prolifera*. The number 83 was assigned to this hybrid acroporid when the Center for Biological Diversity petitioned to list it under the Endangered Species Act, along with eighty-two non-hybrid coral species.[60] The NMFS rejected the petition for *Acropora prolifera* up front, stating that listing this coral as threatened was not warranted because, as a hybrid, it does not constitute a species as defined by the Endangered Species Act ("it is not known to interbreed, and therefore it does not meet the Act's definition of a species").[61] Margaret Miller, who sat on the biological review committee for NOAA, emphasized that the concern was not the artificiality of the hybrid, as in many other cases of conservation management, because in this case the hybridization occurs "naturally." In her words, "Because we have a naturally occurring hybrid, where two parents will cross and create this hybrid in nature, we're not as concerned about the natural hybrid causing genetic issues for the parents."[62] Despite the lineage of this particular hybrid, however, NOAA was unwilling to even consider listing coral 83. It has also been refusing permits requested by nurseries (like the one mentioned by Ken Nedimyer in our interview) to propagate this species for restoration. Miller explained that "we want to be very thoughtful about whether their artificial manipulation might result in a disproportionate number of hybrids."[63]

NOAA ended up removing hybrid coral 83 from the list of species considered for protection. As the next chapter will discuss, hybrids not only contest traditional classificatory schemes, but they also challenge our understandings of the evolutionary process itself. It is no wonder, then, that the legal system, which is concerned with performing its own order as well as with validating the scientific one, either doesn't engage in their protection or altogether hinders it. But as I will show in the next chapter, the hybrid acroporid coral may very well be kicking right back, thereby enacting hope—albeit of a different kind than many have hoped for—for future life in the ocean. Such performances of resistance on the part of unruly corals challenge legal systems to respond, modify, and adjust, exemplifying how corals can affect and even shape—or breathe life into—law.

SIDERASTREA GLYNNI

Another example of the challenges of coral classification—and how they may breathe life into or out of actual coral bodies—came up in my meetings with Zac Forsman. After talking on Skype and e-mailing back and forth, I finally met Forsman face-to-face at the Hawai'i symposium. After the conference, he invited me to check out the new coral nursery he had helped build at the Anuenue Fisheries Research Center near Honolulu. This nursery is part of an effort by the State of Hawai'i to expedite the recovery of the islands' slow-growing corals. Another goal of the nursery is to serve as a coral bank—a backup storage space for a range of imperiled and endemic coral.

The nursery was a surreal site. Large and noisy tanks, tubes, and pipes overshadowed small aquariums containing beautifully crafted corals in all colors and shapes, as well as shrimp, fish, and anemone. While the structure felt industrial, the corals lent it a soft and fantastical touch. Forsman explained that although many of the corals are extremely rare, they have not been listed as threatened because it is unclear whether they are separate species or are part of a spectrum of morphological variability. Nevertheless, his nursery propagates these corals both to learn about them and to prevent their extinction in the wild. Evidently, the hegemonic reach of species classification in coral conservation has not reached this nursery; in the midst of catastrophic pollution and warming, creative forms of activism emerge.

Toward the end of our tour, Forsman told me about the current clashes between coral scientists who use the traditional way of identifying coral species through morphology and those who favor more novel methods of identification that rely on genetic sequencing. Genetic sequencing has rewritten many species classifications in corals, he explained. What appears to be similar from a morphologic point of view might be very different in terms of genetics, and what appears to be different might actually be part of the same species complex. One result of this novel work, he said, is that "we're finding new species all the time. They're often very cryptic." He continued,

> We were also surprised to find that some very different-looking corals can inter-breed and hybridize and make all these diverse forms. To make matters worse, some of them are very plastic and grow in different forms in response to light or water motion. Because there is so much variation, it hides things that are not as exciting-looking, but these cryptic species are actually quite distinct when you take a closer look under the microscope.[64]

Alongside the discovery of new species—and thus the making of potentially new life to be protected under the law—the use of genetic tools has resulted in questions about the validity of a variety of existing coral species classifications. "I actually kind of killed a critically endangered species," Forsman admitted. Although he said this half-jokingly, it was clear that he wasn't taking it lightly. I insisted on hearing the entire story. At the time, Forsman was writing his dissertation and conducting fieldwork on the isthmus of Panama, he told me. On this small stretch of land, he documented five small colonies of *Siderastrea glynni*, a species that was named in honor of coral scientist Peter Glynn[65] and subsequently listed as critically endangered on the IUCN Red List[66] and as a foreign endangered species under the Endangered Species Act.[67]

The IUCN Red List is configured by a highly respected scientific body that evaluates the level of threat for species on a global scale.[68] Although a listing doesn't trigger direct legal consequences and, on its face, doesn't take prospects of recovery or political and economic criteria into consideration, in practice the Red List is widely relied upon as a justification for listing projects by national legislative bodies worldwide (for example, the Center for Biological Diversity relied on the Red List designation in their mass petition to list the eighty-three coral species).[69]

When Forsman genetically sequenced *Siderastrea glynni*, he discovered that they were not at all what they were supposed to be. Instead of rare species, he found corals who were in fact members of a common Caribbean species. How they got to Panama was, at the time, a conundrum. These findings resulted in a 2005 report subtitled "Is *S. glynni* Endangered or Introduced?"[70] Notably, the normative distance between these two categories could not be more extreme: endangerment triggers modes of legal protection, whereas introduction[71] results in either a withdrawal of legal protection or a removal—an extermination, even—of the introduced organisms.

Forsman was uncomfortable with his scientific finding. The coral he sequenced was named after his mentor's mentor, whom he greatly respected. More importantly, Forsman realized that as a result of his research, this coral would now be delisted from the Red List and from the Endangered Species Act, which would translate into decreased levels of legal protection, conservation funding, and public and scientific attention. Despite his training as a geneticist, Forsman expressed his concern that the extreme power assigned to genetic assertions of identity may not be completely warranted. He admitted that while genetic identification holds the promise of resolving old conundrums by perfectly reclassifying coral species, "the genetic work has [in fact] been very challenging and difficult to interpret." He emphasized, finally, that traditional methods for classification are still very important. In his words:

It is hard to separate the coral DNA from the algae DNA that lives in its tissues, as well as from the many creatures that live on or in the coral. My work has been focused on figuring out these genetic puzzles so they can be used as a tool more widely. But we also [still] have to look at the morphology really carefully and do a lot of measurements to understand the microscopic structure. Sometimes we need to grow corals in different environments to learn how plastic they can be. Ultimately, we need to know if they interbreed or not [in order] to determine if they are separate species. What is a unique, separate group? I think we're beginning to understand that, but with corals, even the experts don't know how to identify many coral species. This can be a big [problem], if not one of the biggest, . . . facing coral protection under the Endangered Species Act.[72]

Forsman's depiction highlights how corals often resist classification, challenging the relevant laws that must then attempt to reclassify them so as to enable their governance.

In the wake of Forsman's findings, another scientist proceeded to genetically test the *Symbiodinium* algae living within *Siderastrea glynni*.[73] He found the exact same clonal genotype that had rapidly spread through the Caribbean. This finding provided additional evidence that *Siderastrea glynni* had been introduced and is not a distinct species. "I felt like we probably have a closed case, or the closest we can get to a closed case," Forsman recounted.

Forsman finally met Peter Glynn in person at the Hawai'i symposium and apologized profusely for raising doubts about a species that was named after him. Glynn was hardly upset. In fact, it turned out that he had recently discovered precisely how *Siderastrea glynni* was introduced to Panama. Thinking that the coral skeleton was dead, scientists had used it for a bioerosion study. But it wasn't completely dead.[74] In other words, the scientists themselves had caused the introduction of this species into the Pacific. Glynn published this discovery in the journal *Coral Reefs* on the very same day as his conversation with Forsman, who recalled, "I had so much doubt for so many years, so it was nice to see that case being closed. Hopefully I can name a new species after Peter Glynn."

Curious to see how this story unfolded in the IUCN's listing records, I checked out the Red List's online page for *Siderastrea glynni*. To my surprise, the species was still listed as critically endangered (figure 29).[75] I notified Forsman, who was more surprised than me. Evidently, lists and other forms of legal classification have their own life trajectory and technological lineage that are not necessarily aligned with the scientific ones. In this instance, the

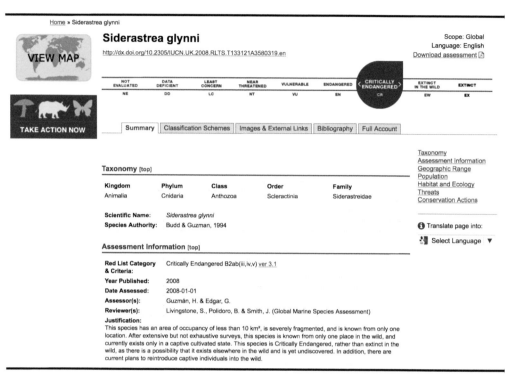

FIGURE 29. The IUCN Red List assessment of *Siderastrea glynni*, screenshot from July 14, 2017. Courtesy of Craig Hilton-Taylor, IUCN.

IUCN listing documents, which constitute a "soft" law in the sense that they are unenforceable despite their strong normative authority,[76] seem to lag behind the science.[77] To quote Foucault: "It is not that life has been totally integrated into techniques that govern and administer it; it constantly escapes them."[78]

QUANTIFYING CORAL: WHY EXISTING METRICS DON'T WORK

The relationship between law and science runs deep, and in the case of coral conservation it plays out in interesting ways. Federal officials are required by the Endangered Species Act to use the best available science to determine whether certain coral species are endangered and thus worthy of protection and recovery efforts. This exclusive reliance on hard science translates into scientific practices of identifying, classifying, tracing, and counting corals and their rates of decline. Is a particular coral species likely to become extinct by 2100? If so, it is proclaimed endangered; if the species is not currently endangered but likely to become so in the near future, it is classified as threatened.

Jennifer Moore and Margaret Miller of NOAA followed the corals' listing process from the petition stage, through the status review, until the final listing. They told me that after examining the four main climate scenarios identified by the Intergovernmental Panel on Climate Change in 2014,[79] NOAA decided to use the most dire prediction of the four for their calculations of endangerment—RCP8.5—which predicts the highest level of CO_2 concentration in 2100. Moore explained this choice: "When we list a species, we are not supposed to use the precautionary principle [i.e., the worst-case scenario]; we're supposed to use what is likely to occur. But because the [climate] experts didn't say which one [of the four scenarios] is likely to occur, all we could do was to make sure we are not underestimating the risk."[80] As it turns out, uncertainty doesn't always have to be detrimental to progressive climate change policy—scientists aren't always "merchants of doubt."[81]

Nonetheless, listing corals remains a daunting task. A pervasive problem with existing coral conservation laws is the inadequacy of the metrics deployed for determining the legal status of coral species. Specifically, one of the central measures used for threat assessment is population size.[82] The assumption is that it's better to have more members of a certain species than to have fewer of them. Margaret Miller introduced me to what she called the "number buffer" assumption, which supposes that sheer numbers can themselves offer a measure of protection. The buffering capacity of corals is high, she told me, as their numbers are many times larger than those of more conventional endangered or threatened species, especially terrestrial ones. So, how should scientists and legislators compare the endangerment of a particular coral species that has millions of existing colonies—take the staghorn coral, for example—with that of the last two remaining northern white rhinos on the planet? How many staghorns equal one rhino? Existing practices of comparing numbers to determine levels of endangerment don't offer much guidance or meaningful information.[83]

Alongside the general problem with sheer number comparisons across the board is the challenge of actually counting corals. The first difficulty that legal administrators and scientists have faced when counting corals is figuring out how to distinguish the individual from the colony.[84] Miller explained that "at certain sites, we can go and sample all the colonies at a site, and it's all the same [genetic] individual. Other reefs and other geographic locations, we can go and sample fifty to one hundred colonies and they each represent a different genetic individual. It runs the entire spectrum."[85] Alison Moulding of NOAA explained the theoretical conundrum that this raises: "What is the [coral] individual? Is it the polyp, the colony, or the genotype? Each has its problems. If you look at a [coral] thicket, you can't necessarily count how many colonies are on that site because it's all one big massive tangle, [and] you can't determine where one starts and the other ends."[86]

Questions regarding the boundaries of the coral *genet*, the scientific term for an asexually reproduced colony of polyps who share the same genetic identity, are becoming even more challenging in light of recent research on symbiosis that complicates this identity.[87] Furthermore, the corals' clonal nature presents challenges to traditional assumptions about how harm and death occur. In particular, the legal term *take*, which is a central tenet of the Endangered Species Act, aims to prevent any harming of listed species by physical injury. Yet this term may need to be deployed differently in the context of corals, whose fragmentation creates new life and is in fact utilized by coral nurseries for restoration purposes.[88] "You can break off a piece of staghorn coral, but you're not killing it, you're just making two colonies," Moulding told me.[89] Immersing ourselves in, and making sense of, coral identity is therefore a good way to think about, and contest, the assumed stability of the individual and the assumed fixity of the law. What concerns me here is how scientific questions about corals manifest in the legal context, which is arguably as constitutive of coral life as the biological and the cultural contexts.

Compounding the difficulties in discerning coral individuals from collectives is the issue of age, which is another common criterion for measuring endangerment. But how does one measure the age of a coral? Ruth Gates, director of the Hawai'i Institute of Marine Biology, reflected in this context: "You could have an [individual polyp] at six hundred years old, but that doesn't mean the living tissue is six hundred years old. We don't really know how those two things relate."[90] Moulding encounters this issue on a regular basis when designing coral recovery plans. "There's no way to calculate the age of a polyp," she acknowledged in our interview. The age of certain coral colonies can be assessed by counting their rings, which resemble rings in trees. "But that's the colony age," she clarified, "not the polyp age." For these reasons, the colony, rather than the polyp, has been used as the scientific and legal basis of

measurement for recovery plans and for determining the details of conservation protection in each case.[91]

Finally, as if challenges surrounding identification, classification, and
quantification are not enough, corals present legal challenges due to the paucity of available historical data about them, which affects the ability of both
scientists and legal practitioners to adequately assess their decline (recall the
earlier "shifting baseline" discussion) and manage their conservation. Margaret Miller explained that "we don't have good historical data. All we have to
go by are the few historical datasets, [which] show a 90 percent decline—
maybe—at a couple of sites. So we're forced to use qualitative information. . . .
We infer."[92] Margaret Moore, who participated in the corals' biological assessment mandated by the Endangered Species Act and executed by NOAA,
summarized the overall difficulties that emerged during that process. In her
words,

> The types of things people typically consider when they're looking at extinction
> risk, like generation time and productivity, are hard when you're talking about
> corals that can live hundreds of years, reproduce both sexually and asexually, and
> are dying at alarming rates. Trying to tease all of that out under a law that was
> written with wolves and whales in mind is a very challenging task.[93]

"We diligently went species-by-species, trying to be consistent between species," Moore told me about the coral listing project. "Our Final Rule document,
in Word format, was over one thousand pages long."[94]

Although NOAA has invested immense efforts into the listing processes
mandated by the Endangered Species Act, the efficacy of these efforts remains
unclear. "The factors that have affected the corals' really bad decline are largely
ones that we don't manage directly," Moore explained, acknowledging that
"the tools that [government agencies] have in their toolbox are not ones that
can fix disease and global warming, or ocean acidification."[95]

INTERNATIONAL CORAL LAW

Harvard law professor Richard Lazarus characterizes climate change as a
"super wicked problem" that "defies resolution because of the enormous inter-
dependencies, uncertainties, circularities, and conflicting stakeholders implicated by any effort to develop a solution."[96] His colleague Jonathan Lovvorn
asserts, further, that "over the last several decades, the traditional domestic
and international environmental command and control regulatory schemes

have repeatedly failed to meaningfully address . . . climate change impacts."[97] He adds: "The voluntary nature of international law, and the notorious difficulties of enforcing international agreements, only exacerbate the already troublesome nature of trying to regulate climate emissions." Despite these challenges, emerging international regimes attempt to protect threatened corals, signifying hope even as the force of law seems less than forceful.

Few international laws directly address coral reefs.[98] Corals received some attention in 1992 at the United Nations Earth Summit in Rio de Janeiro, where participants from 172 governments and more than 2,400 representatives of nonprofit organizations reached an agreement on the Climate Change Convention, which in turn led to the Kyoto Protocol and the Paris Agreement. Corals were addressed in chapter 17 of *Agenda 21*, the United Nations' blueprint for sustainable development. In 1994, the International Coral Reef Initiative (ICRI) was announced at the first "ordinary meeting" of the Conference of the Parties to the Convention on Biological Diversity. Founded by eight governments—Australia, France, Japan, Jamaica, the Philippines, Sweden, the United Kingdom, and the United States—ICRI now counts more than sixty members. In 1994, ICRI launched an ambitious project called the Global Coral Reef Monitoring Network, which aims to increase the scientific understanding of and the public awareness about reefs.[99] Finally, ICRI also joined more than one hundred governments as well as scientific and environmental organizations to declare 1997, 2008, and 2018 as International Years of the Reef.[100]

The Convention Concerning the Protection of the World Cultural and Natural Heritage, more commonly known as the World Natural Heritage Convention, is an international agreement adopted in 1972 by the member states of the United Nations Educational, Scientific and Cultural Organization (UNESCO).[101] The Convention provides yet another mechanism for protecting coral reefs at the international level: the declaration of UNESCO World Heritage Sites, which is intended to unite the world around a shared responsibility to protect natural and cultural places of "outstanding universal value." Of the twenty-nine reefs designated as World Heritage Sites under the Convention, including the Great Barrier Reef in Australia and Tubbataha Reefs Ocean National Park in the Philippines, twenty-seven underwent severe bleaching in 2016–2017.[102] In its 2017 report, UNESCO's World Heritage Centre in Paris cautioned that without dramatic reductions in greenhouse gas emissions, these heritage reefs "will cease to host functioning coral reef ecosystems by the end of the century."[103] Although the report expressed "utmost concern" regarding the impacts of climate change on the World Heritage reefs, some scientists were disappointed that instead of calling for immediate action, it only recommended "further study [of] the current and potential impacts of climate change" deferring other actions until the committee meets again in 2018.[104] This is yet another example of the frustration of scientists with the fundamental inability of international apparatuses to adequately protect corals in the Anthropocene. There is one partial exception in this regard: CITES.

The Convention on International Trade in Endangered Species of Wild Fauna and Flora, or CITES, is a powerful international treaty that requires a permit to trade in vulnerable species beyond national borders.[105] Signed in 1975 and ratified by 183 countries, CITES protects certain coral groups such as black, stony, blue, fire, and lace corals. This protection is quite limited, however. The treaty presumes that trafficking and trade in wildlife and wildlife parts is a primary threat to species survival. Other threats, like pollution, acidification, and warming are simply not contemplated by CITES's regulatory approach. So while it "has teeth," legally speaking, it is also extremely narrow.[106] Furthermore, although such stringent international protections may seem effective for globally protecting corals, some of my interviewees have expressed their concern that the inconsistent listing of certain corals but not others under CITES has, in fact, produced problematic results. While "some of the precious corals that are actually used and globally traded have not been protected through CITES," deep-sea coral scientist Murray Roberts told me,[107] other widely found coral species are granted CITES protections, which makes it more difficult to research them.[108] Law materially affects corals, and thus scientists must both contend with, and negotiate, its rule.

A CONSTITUTIONAL RIGHT TO CORALS

In the United States, the problem with international law and the lack of climate change statutes at the federal level have engendered creative litigation strategies. In *Foster v. Washington Department of Ecology*, eight schoolchildren from Washington sued their state government, asserting their "inherent and fundamental rights to a healthful and pleasant environment."[109] To everyone's surprise, the judge in that case ruled, albeit *in dicta* (i.e., not as a binding precedent), that "global warming causes an unprecedented risk to the earth," that the state constitution bestows environmental rights, and that the public trust doctrine extends to the atmosphere. Six months later, the court ordered the Washington State Department of Ecology to finalize the climate change rule by the end of 2016. The department complied with the court's order, issuing the final Clean Air Rule, which establishes new greenhouse gas emissions standards for "certain stationary sources, petroleum product producers and importers, and natural gas distributors."[110]

Building upon this success, the nonprofit that litigated *Foster*, Our Children's Trust, convened a separate group of plaintiffs to file a complaint at the federal court level. This, they hoped, would result in a decision that would set an authoritative precedent covering the entire court system. In 2015, twenty-one young people, including nineteen-year-old Kelsey Juliana, filed a complaint against the United States government. *Juliana v. United States*[111] seeks to constitutionalize climate change policy.[112] The complaint asserts that the government's actions that cause climate change violate the young generation's constitutional rights to life, liberty, and property, and also fail to protect essential resources in the public trust. To quote from the complaint:

> Youth Plaintiffs have a substantial, direct, and immediate interest in protecting the atmosphere, other vital natural resources, their quality of life, their property interests, and their liberties. They also have an interest in ensuring that the climate system remains stable enough to secure their constitutional rights to life, liberty, and property, rights that depend on a livable future.[113]

One of the plaintiffs, Journey Z. from Hawai'i, drew a direct reference to coral reefs, their endangerment, and their centrality to her personal life and well-being. Based on evidence collected from coral scientists, the complaint states that "at a prolonged 450 ppm [parts per million[114] of carbon dioxide] level, coral reefs will become extremely rare, if not extinct, and at least half of coral-associated wildlife will become either rare or extinct. As a result, coral reef ecosystems will likely be reduced to crumbling frameworks with few calcareous corals remaining."[115]

The plaintiffs achieved an early victory: on November 10, 2016—one day after the election of President Trump, who, in 2012, referred to climate change as a "myth" propagated by the Chinese—Judge Ann Aiken of the Ninth Circuit denied the defendants' motions to dismiss, recognizing the possibility of an environmental due process claim. In the judge's words,

> Where a complaint alleges governmental action is affirmatively and substantially damaging the climate system in a way that will cause human deaths, shorten

human lifespans, result in widespread damage to property, threaten human food sources, and dramatically alter the planet's ecosystem, it states a claim for a due process violation.[116]

Julia Olson, executive director of Our Children's Trust, explained the power of the constitutional argument in this case: "When you're dealing with an emergency of apocalyptic proportions, where human survival is threatened and civilization is threatened, then you have to step back to fundamental law: constitutional or human rights laws."[117] In other words, there are laws and there is *the law*. "We won't solve climate change with *laws*," Olson told me, "but [we will] with *the law*." Such a capacity to go "back to the roots" is what gives her hope. Focusing on the founding fathers' views on nature, Olson highlights that "these men were farmers, with deep connections to nature. They understood that their ability to grow their own food, and to have access to their own water, was the basis of their liberties." She continued: "It's now up to the courts, at this very late date, to step in and provide a check on the laws that are destroying our natural systems, and go back to *the law*, the fundamental place that we should be working from."[118] The *Juliana* complaint is still making its way through the federal court system. In March 2018, the Ninth Circuit Court of Appeals rejected the Trump administration's "drastic and extraordinary" petition for writ of mandamus to dismiss the complaint without trial. The trial date will be announced shortly.[119]

EXECUTING THE CROWN-OF-THORNS STARFISH

In discussions about the relationship between law, science, and the environment, one tends to focus on law's affirmative projects of protecting imperiled species and habitats. In the final sections of this chapter, I would like to explore another aspect of the law that is perhaps less widely discussed but certainly not less common: killing, or, in Foucauldian terms, "making die."

The crown-of-thorns starfish (*Acanthaster planci*, or COTS for short) is an unusually large starfish with seven to twenty-three arms, all bristling with spikes. Native to the Indo-Pacific region, the starfish has one of the highest fertilization rates recorded in any invertebrate: a large female is capable of producing up to sixty-five million eggs over one spawning period.[120] The starfish feed on corals. In fact, they used to perform an important role in increasing coral diversity, as they tend to eat the fastest-growing corals, such as staghorns and plate corals, thereby allowing slower-growing coral species to form colonies. Yet changing conditions have resulted in starfish population numbers spinning out of control. In a 2012 report, Australia's Great Barrier Reef Marine Park Authority (GBRMPA) suggested that outbreaks of this venomous invertebrate have been posing a significant threat to the Great Barrier Reef. Coral cover has declined by about 50 percent over the past thirty years, the report stated, deeming the crown-of-thorns starfish responsible for almost half of this decline.[121] The report also included a few daunting images of the starfish sucking the life out of corals.[122]

As a result of the report's findings, the Australian government published an instruction manual for "culling" the starfish. In "a program for the taking of animals, which pose a threat to the use and amenity of a particular area," the government advised that "the best practice method for undertaking COTS control is to use a modified drench gun to inject the starfish, using either the single-shot bile salts or the multi-shot sodium bisulfite method. These injection methods minimise the risk of breaking corals, and are safer than manual removal."[123] At a cost of three cents per injection, the law thus delivers a death sentence to this animal, in turn celebrating this death as a way to make corals live.

David Wachenfeld, director of coral reef recovery at GBRMPA, admitted in our 2016 interview that it is highly unlikely that conservation efforts will stop the crown-of-thorns starfish outbreaks. Instead, the rationale behind the organized killing of this animal in very particular locations is that after the outbreaks are over, these sites would serve as a "natural nursery" for replenishing the reefs of that area. GBRMPA decided to prioritize twenty-one coral reefs for this purpose. "What we're hoping to achieve by tactically controlling the COTS is to create reefs where there is still good coral left that will serve as reproduction centers that will kick-start the next recovery phase," Wachenfeld told me. GBRMPA's main objective is not to kill starfish, he insisted, but rather to maintain the capacity of corals to reproduce. The problem facing the Great Barrier Reef is the accumulated impact of bleaching, cyclones, and crown-of-thorns starfish, he added. "We can't stop bleaching and we can't stop cyclones," he admitted. "But we can do something about [those] starfish."[124] The impetus to take action, coupled with the perceived inability to mitigate climate change, has resulted here in the identification of a more practical, and vulnerable, target,[125] with the added bonus of shifting some of the blame from the human perpetrators to the starfish.

KILLER ROBOTS

In a recent turn of events, coral managers have been enlisting nonhumans in the war against the crown-of-thorns starfish and for the protection of corals, thereby instantiating yet another displacement of humans from the scene of ecological devastation. In 2015, researchers from Queensland University of Technology conducted final trials for a robot designed and trained to administer the lethal injections to the starfish: the COTSbot (figure 30). The COTSbot would be sent out to the reef for up to eight hours, delivering more than two hundred lethal shots a day. "It will never out-compete a human diver in terms of sheer numbers of injections but it will be more persistent," COTSbot designer Matthew Dunbabin said,[126] explaining that the robot could be deployed in conditions that are unsuitable for human divers, such as nighttime and inclement weather.

Currently, the robot must receive the approval of a licensed human for every lethal injection. However, Dunbabin hopes to soon move away from human verification and give the robot "as much autonomy as we can."[127] The COTSbots photograph their targets, he told me, and in cases of uncertainty, they require human verification before injecting the crown-of-thorns starfish. Such instances of uncertainty provide further training opportunities for the COTSbot: after verification, the photo is added to the robot's arsenal of images to help it better distinguish between killable starfish and livable coral.

While the COTSbot is currently activated by experts, Dunbabin highlights the potential for its activation by "citizen scientists." To achieve this, he plans to design an easily accessible control system that will deploy a smartphone to manage the robots across the ocean. In these instances, the robots would not be used for culling, he assured me, but only to monitor starfish populations and map the ocean floor. If his vision materializes, a fleet of robots will roam the oceans, rendering the undersea world visible to humans, albeit circumscribed by the singular perspective of the killer bot. Such surveillance technologies could potentially be deployed by any interested party, raising questions about the legality of these projects across jurisdictional boundaries. Ultimately, Dunbabin hopes that his robots will be like drone aircrafts: "If you can have millions, that would be awesome."[128] At the same time, he is careful to emphasize that while the military drone is designed to kill, the

FIGURE 30. From the website of Queensland University of Technology (QUT): "QUT researchers Dr. Matthew Dunbabin and Dr. Feras Dayoub have successfully completed field trials of their COTSbot robot, which they've proved can navigate difficult reefs, detect COTS and deliver a fatal dose of bile salts—all autonomously and precisely." Credit: Richard Fitzpatrick for QUT (used with permission).

COTSbot is made to save. "We haven't tried to hide anything," he told me. "This is all that it does."[129]

Animal rights groups have not protested against the culling of the crown-of-thorns starfish, and the institutional review boards that authorize such research projects in university settings have also been silent on this front. When I asked him about this, Dunbabin explained that

> the starfish is an invertebrate, and there [are] only a couple of invertebrates that are considered animals in the eyes of ethical review boards. One is the octopus, and I can't remember what the other one is. Maybe the coral. So it was very easy to get the ethics committee's approval for this project. Dolphins would be a different story.[130]

Although they are new to the project of restoration, robots and other machines have already been used extensively to map, survey, collect data, and monitor the ocean environment. For the past twenty years, deep-sea biologist Murray Roberts has been using remotely operated vehicles for his research. My interchapter interview with him, following this chapter, expands on these experiences.[131] While such machines are invaluable for gleaning any kind of knowledge about the deep sea, I would argue that the growing use of robotic technologies in the ocean also contributes to the problematic sense that a technological fix to the world's problems is possible, and even under way. Finally, let me highlight again that while robots bring humans virtually closer to this wet and largely dark space, they also serve as displacers: shifting human responsibility, if not human agency, away from the scenes of violence.

BACK TO THE BREATH

The importance of corals in the intersection of climate change and law is critical.
—Julia Olson, 2017[132]

The project of managing coral is, more than anything, a coralation of law and science. This chapter has pointed to the coproductive as well as the unproductive relationships between lawyers and scientists as they engage in coral conservation. Although the two groups share similar goals, the perceived and real differences between them have often been a source of frustration for both sides. Such differences—in risk tolerance, speediness of procedures, and scale of operations—have played out in various ways throughout this chapter, revealing the awkward dance that law and science tend to perform with one another: when the law steps three paces forward (for example, when legal administrators offered to assign extra protections to coral species through their uplisting), science steps two paces back (the scientists adamantly refused this offer); and, vice versa, when science steps forward (for example, when geneticists proved that an endangered coral species was actually not a species), the law steps back (by neglecting to amend the listing status of this coral).

Yet, amid the differences and tensions, this chapter has also provided examples of successful collaborations between coral scientists and lawyers. For example, Julia Olson of Our Children's Trust, who filed the climate change lawsuits on behalf of young people, emphasized that, in her experience, coral scientists are not like any other conservation scientists: "You would never hear a coral reef expert say that two degrees Celsius is a target that would protect corals. It's just plainly known [to them] that if we go there, we may not have corals that survive. By contrast, the land-based climate scientists don't really look at targets in that way." Using coral science, Olson argued in the lawsuit that we must "draw a clear line in the sand that if we're at a rate of above 350 ppm, we are in a danger zone for corals."[133] Corals and their scientists have thus been vital for *Juliana v. United States*.

FIGURE 31. Zac Forsman of the Hawai'i Institute of Marine Biology explained: "This is *Montipora dilatata*, [which used to be considered] one of the rarest corals in the world. Only a dozen or so colonies could be easily identified. [As it turns out,] this coral looks really distinct, but it might not be a separate species from more common corals, according to some genetic and morphological data. My genetic work ended up preventing this coral from being listed under the Endangered Species Act. I have mixed feelings about this result, as I have with the case of *Siderastrea glynni*. I have never been one for doom and gloom, but seeing these corals being lost to coral bleaching in 2014 was the first time I cried underwater. This illustrates the big problem about using the Endangered Species Act for corals: we don't really understand coral species. Many of them may hybridize, or they may be much more polymorphic and variable than anyone thought. I'm working on trying to answer this question" (e-mail communication, June 5, 2017). Courtesy of Zac Forsman.

But while creative and dynamic interpretations of the law such as those offered by the *Juliana* case give some hope for coral management, one must at the same time acknowledge that the myriad coral listings, petitions, lawsuits, mitigations, and permits have so far failed to protect corals in their rapidly changing ocean environment (see, e.g., figure 31).

As I've repeatedly emphasized throughout this book, corals force us to take a long, hard look at how we think about the world. In this chapter, the corals have exposed the anthropocentric bias at the heart of law, demanding innovative approaches from us if their decline is to matter for how we manage this planet in the future. The chapter has also detailed the nuanced ways in which legal administrators, equipped with words and paper, stretch, bend, and lengthen legal norms to fit the particularities and peculiarities of coral life—thereby breathing life into corals. Their imperative is to make the coral visible to the law, and they have been using the legal and scientific language of endangerment for this purpose. At the same time, legal practices also kill. I've shown how such lethal legalities take place in the cases of coral hybrid 83, *Siderastrea glynni*, the crown-of-thorns starfish, and *Montipora dilatata*.

Because humans are the dominant species in the Anthropocene, coral life is very much dependent on our priorities, our choices of protecting and conserving certain species rather than others, and our existing legal and administrative systems for classifying, regulating, and governing life. If we are to truly see and protect corals, then, changes to the current regulatory regimes must be considered that take their nature into account. The task of rethinking coral law and making it more coralated is especially urgent at this time, when coral ecosystems are disappearing at a dizzying pace. Corals can serve as our guides on the journey toward a much-needed alliance between science and the law and, specifically, toward recognizing less-like-us beings in alien-to-us environments who nevertheless depend on us for their continued survival.

The Cinderella of Corals

An Interview with J Murray Roberts

J Murray Roberts is a professor of applied marine biology and ecology in the School of GeoSciences at the University of Edinburgh, Scotland. His main research focus is deep-sea, or cold-water, corals. I interviewed him by Skype on January 27, 2016, and then over breakfast at a café in Edinburgh on April 20, 2016—in between two of his long excursions into the deep sea. I contemplated joining him on one of these excursions, but after finding out that this might require jumping out of a helicopter into the ocean, I thought better of the idea. I was interested in hearing from Roberts why deep-sea coral reefs were not really recognized as reefs by many coral conservation scientists (in his words, they are "Cinderella species") and how the material differences between deep- and shallow-water corals play out in terms of regulatory regimes and in the relationship between scientists and lawyers in particular. In his focus on the Cinderella of the coral world, rather than her more glamorous and visible tropical sisters, Roberts's narrative is an outlier. But then so are all of the other interchapter interviews, in significantly different ways.

IB: Many coral scientists don't include deep-sea corals in their inventory of corals of the world. For example, "Charlie" Veron's comprehensive book *Corals of the World* doesn't include deep-sea corals at all. How do you feel about that?

MR: My particular specialty is deep-sea corals, but I work with tropical systems, too. It's surprising how similar they are, in some ways. They are corals, number one. *Coral* is not a very specific biological term. It's like if I said I saw a bug crawling up the wall—you wouldn't know if I was talking about a beetle, an ant, or a cockroach—they're all insects. So it's like that. Just as there are lots of insects, there are also lots of different types of coral. But if you were to add them all up, there are more coral species [living] deeper than fifty meters than [there are living] shallower than fifty meters. The weird thing is that although the world focuses on shallow tropical corals, for all kinds of good reasons, more than half the diversity, really, is in deep waters.

IB: What do we know about deep-sea corals and how are they different from tropical corals?

MR: If we're talking about framework-forming corals that make deep-sea reefs, there are only about six or seven species we know of—very few species compared with many tropical coral reefs. If we're talking about deep-sea corals in general, not just those that make reefs, then there are loads and loads. Photosynthesis doesn't happen in the deep. In shallow tropical reefs, coralline algae bind the reef together—and that doesn't happen [in deep-sea reefs] either. But what does happen is that these corals have adapted to growing frameworks that trap sands and muds that move through the areas. The currents sweep them through and there's a virtuous cycle whereby the corals trap the sediment. If the polyps keep growing, and they can keep pace with the sediment, then, over time, the virtuous cycle produces a mound that grows off the seafloor (figure 32). That's how these mounds can grow even two hundred meters off the seafloor. How do they do this? Well, the better the coral can feed, the better it can grow, the more successful it will be in reproduction, and it will produce a virtuous cycle going around. So there are similarities and differences, key differences, between the two systems.

IB: How did scientists find out about these mounds?

MR: The work we've been involved in for the last twenty years brought to the fore how elaborate deep coral habitats are. And this is new, because in the past we didn't have the technology to work in those areas. There are areas, even off the coast of Scotland, where deep coral mounds grow two hundred meters above the surrounding seafloor, all at a depth of six

FIGURE 32. Murray Roberts explained that "the Logachev coral carbonate mounds are giant seabed mounds constructed over hundreds of thousands of years by the complex interplay of coral framework growth and the trapping and baffling of seabed sediments" (e-mail communication, July 3, 2017). This photo was shot by a remotely operated vehicle on the 2012 Changing Oceans Expedition, where Roberts was the principal scientist. Courtesy of Murray Roberts.

hundred to eight hundred meters. And those coral mounds extend for tens of kilometers—they're like big chunks of the Great Barrier Reef, only they're growing in the cold and dark North Atlantic waters. They were discovered in seismic surveys by the oil industry back in the 1990s. Previously, we knew of deep-sea corals from fishing records: trawlers have run over these areas and brought them up in their nets, so lots of records from there. There were also historical surveys, like the pioneering oceanographic expeditions of the late nineteenth century. So the scientific community had some awareness. But when the geophysics and the [oil and gas] industry came in, we started to get a much more accurate sense of the real scale at which these corals can grow.

IB: What are some of the challenges of researching deep-sea corals?

MR: When you're dealing with something eight hundred meters down, before you can do anything you need a research vessel, ideally equipped with both mapping technology and good camera systems that can be lowered to the seabed to discover what's there. Whenever we can, we try to use remotely operated vehicles (ROVs) to explore and sample deep seabed habitats. In the past, we've used manned submersibles, but that's a lot less prevalent these days. In the United Kingdom, we use a ROV, which has cameras, lights, manipulators, and a tether all the way back up to the ship (figure 32). This thing launches and down it goes—and we sit with the pilots in the lab or in the container and we direct things from there. We can sample, we can survey, we can visualize.

ROVs are the Rolls Royce of deep-sea technology, and they're only available occasionally, even in the most well-developed scientific economies. They're really expensive. From that you go to simpler technologies. Grabs that can sample the seabed, or cameras that you just dangle from a wire all the way down, or what we did in the Victorian era and we shouldn't do anymore, which is dredge or trawl and just scrape things back up and dump them on the deck and see what we find. The kinds of offshore science-class ROVs will cost a couple million dollars to buy. . . . [So] what we do in Britain is we have one world-class vehicle that is based in Southampton and operates nationally, and we all bid. If it comes through, we get that vehicle for around thirty days at sea. Those thirty days will cost the British taxpayer about 400,000 pounds. Some of the reasons this toolkit is so expensive is that it's deep sea. Picture the research ship I'm on: it's one hundred meters long and weighs five thousand tons and it can deal with twenty-meter waves. The vehicle has to be strong enough to bounce off the side of it if something goes wrong. So all of this inflates the cost dramatically. Recently, the autonomous underwater vehicles are becoming much better. They'll never completely replace the human element though. They do a different job.

IB: I never heard about deep-sea corals until I started the research for this book. How do you explain the public's lack of awareness on this front?

MR: It's strange how we focus so much on the heavens and yet we don't look at our own earth. People don't get the chance to go and see it for themselves, so it's not in their awareness. I have had the chance to be in manned submersibles a few times, and I love it. But the sea is jolly uncomfortable,

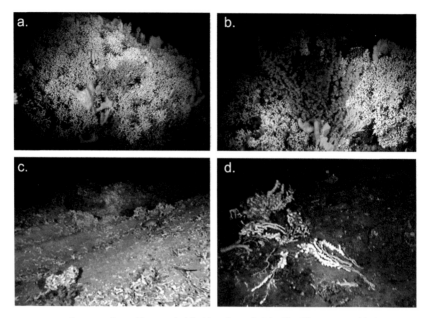

FIGURE 33. A comparison of untrawled (a, b) and trawled (c, d) cold-water coral habitats in Norway. The reef frameworks formed by *Lophelia pertusa* are reduced to rubble, and large gorgonians, such as *Paragorgia arborea*, are broken apart in the trawled areas. Images a and b were taken in 2005 in Stjernsund, courtesy of JAGO-Team, GEOMAR Kiel. Images c and d were taken in 1999 in an area close to Iverryggen, courtesy of J. H. Fossa. The images are from Murray Roberts et al., *Cold-Water Corals: The Biology and Geology of Deep-Sea Coral Habitats* (Cambridge University Press, 2009), 334 (reprinted with permission).

you know? You can't rely on it, and people don't like being seasick. All those things become physical barriers. You get launched in the sub and you get tossed around all over the place, and a lot of people would find that pretty horrible. Maybe one day we can dial up these observatories and look at the pictures that are coming in.

IB: What are some of the threats that deep-sea corals are facing today?

MR: The southern North Sea was trawled in the early nineteenth century. By midcentury, they were trawling it several times a year. By the early twentieth century, the northern North Sea was trawled down to one hundred meters' depth. Now, some of the North Sea is trawled over three or four times each year. Many of the fishing grounds [in particular] were trawled frequently, so the ecology has completely changed. Decades later, we still see the smoking gun—the nets lost on the seafloor, the damage from the machinery. Many areas of the continental shelf in the North Atlantic, where deep corals and sponges are found, are [still] open to trawling— and show the scars.

IB: Have there been any attempts to restore deep-sea corals damaged by trawling?

MR: There's been massive concern, quite rightly so, over the damage caused by trawling (figure 33). So if a big trawl net goes through a coral area in the deep ocean it smashes it down. That's where restoring it is practically impossible—although if you could have an army of robots working at those great depths, maybe you could do some work that would enhance restoration. If you're restoring a tropical coral reef, a practical solution is to send people in; they will do the job very effectively. But in the deep ocean that is absolutely impossible. So the robots have this tremendous role to play.

My wife, Lea-Anne Henry, came up with the coralbot idea. She challenged the engineers at our university to figure out a way to pick up a coral that had been broken from the reef and turn it the right way up so that it restores the ecosystem. They did lots of work. They didn't end up raising the kind of money they needed to develop it, but they worked out the image-processing algorithms that the coralbot would use. There will be papers coming out on that very soon. Lea-Anne had a note out in *Nature* about the restoration of the deep sea, authored with a number of other people. The robotic age is coming, and it is critical for the deep ocean.

IB: Does climate change have any impact on deep-sea corals?

MR: Deep-sea corals are much more vulnerable to climate change than one would think—and climate change has really driven our research agenda over here for the last ten years.[1] The primary concerns in the deep ocean relate to [both] ocean acidification and the warming of the seas. Over 90 percent of the planetary warming caused by global warming has been absorbed by the oceans. The deep ocean is now warming, holding less oxygen, and global ocean current patterns seem to be changing. These changes will have profound implications for all ocean ecosystems—but the exact changes are really hard to predict because these ecosystems are so massively complex. Ocean acidification works from the top down, because the CO_2 dissolves in the surface layers and mixes. But actually when you go deeper, the chemical environment for corals becomes harsher. Tropical corals grow in waters that are supersaturated with calcium carbonate by three or four times. When you get to the kind of water masses that our corals are growing in, it might only be just saturated or maybe saturated by two times, sometimes less. You don't have to make [the water] that much more acidic to tip it over the edge and make it corrosive. There's a natural layer in the oceans above which corals can grow because the water is supersaturated for calcium carbonate, but underneath [that layer]—it's undersaturated. And that horizon is narrowing because of all the CO_2 we are adding into the system.

So I would argue that the deep corals are more vulnerable to climate change than shallow ecosystems and that they are seeing that change faster and have less inherent capacities for adaptation. That's actually what I'm researching now: What is the deep-sea corals' inherent adaptation ability? Could they deal with this in the future? We just don't know. We have looked at the implications on present-day corals, and we found that they become very, very weak. Their skeletons and the little crystals that make these skeletons, which are normally lined up and organized, all become disorganized in scenarios that we will definitely see more of by the end of this century. This weakens the living coral colonies. If the dead coral skeletons that give deep reefs their structure are exposed to corrosive seawater, as predicted in many areas by the end of this century, they will start to dissolve.

In the deep sea, the corals are also limited to quite narrow temperature windows. When we've looked at their physiology and we've put them at the upper levels of their ranges, their metabolic rates went up many times. We just don't understand too much about how the ecosystem will respond. In fairly simple terms, if enough food keeps coming down to these places, maybe they can sustain that metabolic activity; but if the food supply changes, as it probably will, we are just not sure what will happen, but we think they will simply starve and run out of energy.

IB: Could you tell me a little bit more about the feeding mechanisms of deep-sea corals? How do they survive in what seems like such a deprived environment?

MR: Deep-sea corals live in darkness. There's no photosynthesis going on. You can take the symbiotic algae found in many tropical corals out of the equation. They're not present. What happens instead is that, like most corals, deep-sea corals feed from the water column: they are adapted and selected to capture food that originates in well-lit surface waters and is transported down. And there's been a whole area of active research in the last decade on these food-supply mechanisms. We've known since biologists first started looking at these things that they grow in areas where the currents are locally accelerated. Within the ocean there are internal waves, just like on the surface. Where those waves break—when they hit the continental shelf edge or an offshore bank—this promotes turbulence and mixing and draws down food from the surface. This explains why these corals grow in such abundance where they grow. It's all about food, but it's food from the upper reaches of the water.

IB: Could you tell me about one specific species of deep-sea coral, or one particular habitat? Maybe one that you have been researching, or is your favorite for other reasons?

MR: *Lophelia pertusa* is the most abundant of the deep-sea corals anywhere in the world (figure 34). Another habitat that we haven't talked about too much but that is critical for deep-sea corals and deep-sea fish is offshore seamounts. These are extinct underwater volcanoes, very mysterious places. The corals really proliferate on seamounts at certain depth bands, often related to internal wave dynamics. Geological processes led to the formation of the volcano. The corals came later, when [the volcano] was already an extinct structure. . . . You end up with whole zones of deep-sea corals and these are the bits that people have only just clued into and discovered. Again, [this is] a place where fishing has caused a lot of damage.

IB: What role has the deep-sea coral played in ocean conservation?

MR: Deep-sea corals have been the poster child for deep-sea conservation and, specifically, for the conservation and sustainable use of marine biodiversity in areas beyond national jurisdiction, which are most of the planet.[2] All the debates happen through the United Nations [UN], because of the UN Convention on the Law of the Sea, or UNCLOS. We are reaching a very interesting time now: there's been a whole decade of careful, hard slog,

FIGURE 34. According to Murray Roberts: "This image of the *Lophelia pertusa* coral colony and the tusk fish *Brosme brosme* was taken from the northwest Rockall Bank at a depth of 220 meters. *Lophelia pertusa* is one of the most vulnerable species to global climatic change; it's also very vulnerable to damage from bottom trawling. I had no idea when I was finishing up my Ph.D. on a Scottish sea anemone that I'd end up devoting the next twenty years of my life to studying a deep-sea coral that no one had ever heard of. This led me to create the first website dedicated to deep-sea corals, which I named *Lophelia.org* to bring this coral into public awareness. I also made a short film about it with David Attenborough" (e-mail communication, July 3, 2017). Courtesy of Murray Roberts.

back-room negotiating, as well as front-room negotiating. This long process is now culminating with the UN discussing whether we need a whole new legal regime to manage marine biodiversity in areas beyond national jurisdiction. But while we are getting better at how we manage these things and how we put out the plans, we still lack much of the critical information. [For example,] we don't know how the corals are connected from one area to the next, certainly not from one ocean basin to the next. We're now starting a new generation of projects here in Europe, which are going to work on an ocean basin scale for the first time and try to tie those big questions of ocean exchange and turnover and variability with connectivity of ecosystems on the deep seafloor. Those who have to put the conservation policy measures in place are the policy makers who often have a legal rather than a scientific background. In terms of the legal side of conservation on the high seas, corals are central because people are concerned about the physical damage of bottom trawling. And that has unfolded over [the last] ten or fifteen years now.

Legally, these areas are defined through UNCLOS as the common heritage of mankind. As such, they belong to everybody. I read this stuff from the UN and I can't get through it; I mean, it's Article number twelve-slash-three blah blah blah—by the time I've read it, I've fallen off my chair. And that's me wanting to understand it. There has been a debate for the last several years, which hasn't received much publicity, about areas beyond national jurisdiction. Fisheries have been the driving force of change in the high seas, really, for the last twenty years, maybe even more. But now there are the emerging issues of climate change, acidification, and new activities like deep-sea mining. Until very recently, it's been more science fiction than fact. But now we're looking at the deep ocean as a potential new source of minerals, and the oil and gas industry has developed subsea technologies that make this possible. There are companies now fully capitalized and ready to start exploratory work, but no exploitation licenses are being granted by the International Seabed Authority, only licenses to explore. The question is, how do you possibly deal with the impacts? To be honest, the impacts are so severe [that] the areas that are mined are gone, they won't be there anymore.

IB: What's your take on deep-sea mining?

MR: I think that it's fundamentally unfair. The mineral-rich areas are situated in international waters—and we have agreed through the UN that they should be managed for common heritage. But because nobody is being provided with information about them and there is no awareness about these areas, it's quite possible for interest groups to go in and liquidate the assets. And that's what has been happening to deep-sea fish in many areas around the world. Quite small interest groups—be they deep-sea fisheries or, now, deep-sea mining companies—are going into areas and brokering deals to harvest. Actually, the fisheries just went in and took the resources without even brokering deals. They were subsidized by very wealthy nations, to the tune of many, many millions of dollars. At the end of the day, deep-sea fisheries have contributed about 0.25 percent of fishery input into already-developed economies like those here in Europe—deep-sea fish aren't a vital part of our food security. It's quite similar to the tragedy of the commons—it's the same way of thinking.

Deep-sea corals are helpful in this context, because when they break they provide physical evidence of the damage to these otherwise invisible common heritage sites and they then become a focus of interest. But it's worrisome that we don't necessarily think enough about what are perhaps less charismatic components of the deep sea: the deep-water sponges, for example. People don't get as excited about sponges, sediment-dwelling communities, in the same way they get excited about corals. I even hear people referring to the deep sea as a desert, because they don't see anything living there. Well, there's lots of life, it just happens to be living in the mud, so you can't see it. But if you carefully sample the mud and sieve out the animals, you can identify the community that has evolved to live there.

IB: So where are things right now in terms of UNCLOS and the regulation of the high seas?

MR: It's kind of up to the delegates and the parties to UNCLOS to decide where it goes from here. They are currently meeting in information-gathering and assessment sessions at the Preparatory Committee—or PrepCom, as they're called—to collectively plan what the UN should do next. The Convention on Biological Diversity, or CBD, has also taken a lot more interest in oceans in the last decade or so. A major catalyst for that was the damage of deep-water trawling and the growing concerns of marine ecosystem damage around the world. When that became obvious, the issue gradually went higher up in the political agenda. And some of the big NGOs have a lot to do with that. They are now asking questions like: If everyone owns the seabed in international waters, who is responsible for protecting it? Who is responsible for restoring the areas beyond national jurisdiction? And how do we actually finance the protection of the deep sea? Some have suggested using principles such as "the polluter pays" and trying to get industry and other sectors more involved. We've been providing some of the scientific input to these discussions. So it's become this intersection of science, law, and policy.

IB: What are some other aspects of this intersection between science, law, and policy?

MR: Small-island developing states are also becoming more and more aware of their rights. Ninety-eight percent of Belizeans voted against off-shore drilling and banned it. And they don't trawl, either. Belize cut the cord, realizing that should an accident ever occur, their tourism would be negatively impacted. The CBD is just watching all this happening and realizing they have to do something. And it's the last ten years, really, that have seen a big change and higher levels of policies addressing deep-water issues. They've come up with various ways of breaking the problem down and thinking, "Well, we've got the whole ocean out there, [so] how do we look at certain areas to see if they're particularly vulnerable to damage?" There are different ways that organizations have done that. The Food and Agriculture Organization of the United Nations put forward a resolution to conserve Vulnerable Marine Ecosystems (or VMEs), which are systems that have certain properties of long growth, slow reproduction, fragility, [and] vulnerability to damage. That was recognized in a couple of UN resolutions. The CBD then initiated a process, which is still ongoing, that defines Ecologically or Biologically Significant Areas, or EBSAs. EBSAs simply define the characteristics and criteria with the best available science—there's nothing about the management. And that's where UN bodies started to enter into difficult situations with their contracting parties. Their parties will say, "Hang on, that's our seabed. You're not telling us how we look after that, that's our affair, thank you very much." So you have to be very careful. . . .

IB: I don't really understand the difference between EBSAs and VMEs. Could you please clarify?

MR: The EBSAs and the VMEs are jolly confusing because there's a lot of overlap. The criteria are similar, so everyone could be forgiven for mixing them up. And VMEs get tricky because some people insist it's only a VME if it's vulnerable to trawling, [and that] you can't use [the designation] VME if it's going to be vulnerable to oil and gas. It's a legal argument, because it was defined by the Food and Agriculture Organization in relation to the fisheries' impacts on trawling—so it's about that. A lawyer's view is black and white, isn't it? You're either guilty or not guilty, it's either this or it's that. [But] the ecosystem doesn't really care if it was damaged by trawling for fish or by a plume from deep-sea mining. The coral doesn't really care. So the

CBD decided they needed another mechanism, another broad set of characteristics that was more all-encompassing to work through their Convention.

IB: How about areas *within* national jurisdiction—are we seeing more protections there?

MR: Basically, the areas of territorial seabed usually run out two hundred miles off the extended continental shelf. Many claims over these territories are disputed and they're all running through the UN as well. When I checked last year, marine protected areas were only 2.8 percent of the ocean—and the word *protected* can just mean a "paper park"—it doesn't necessarily mean that there is no fishing or no oil and gas—it just means that it has some level of prioritization or its been identified on something like the Red List. We're all working to ensure that at least 10 percent of our territorial waters are protected.

The goal [for the deep sea] was also 10 percent. The thing is, [that] many countries rushed to protect a particular site because it was one of the first-ever deep-sea coral finds of an area. And they just went ahead and protected it, which is great, but it's not actually representative of what those sorts of ecosystems can look like. There's so many others we would have picked as scientists. But in their rush to achieve those 10 percent, they have been designating areas with no management and also without the science to say that was a good choice.

IB: Are you coming up with alternative ways of managing the deep sea?

MR: Have you come across Horizon 2020? This is the European Commission's current research funding framework. We're leading a large deep-sea project. It's called ATLAS and it is going to create a transatlantic ecosystem assessment and management plan for deep corals and sponges and other ecosystems in the entire Atlantic. It's a big project lasting four years, with funding of over nine million euros. It has twenty-four different partners, fifteen associate partners, including NOAA and the Department of Fisheries and Oceans up in Canada. It's a real collaboration—a complex beast of a project.

ATLAS is going to develop a new management plan for the North Atlantic so that policy makers can use this plan and adapt it if necessary. We're basically building up a new way to have marine species management.

The challenge at the moment is [that] management measures are put in place for certain sectors. You can actually put values on different ecosystems in the North Atlantic, and use that to help decide what happens if we lost that site. In the scenarios we're going to run with we'll say: "Okay, 50 percent of this reef is gone, do people value that and if so, how would that affect industry, how would it affect policy?" We use a lot of [predictive] modeling. We [input] information on biodiversity, richness, [and] hotspots, and then use another input from genetics. How connected are these things? Maybe I can wipe out this area but it's going to be resupplied by another area? The next layer is socioeconomics: How do people actually value this sort of stuff? We jam all five of those layers in together—oceanography, food web, biodiversity, genetics, socioeconomics—and then we overlay something else: the footprint of human activities. Now you've got a map that shows everything, and that's the concept. We're trying to operate climate change scenarios for the next fifty years, because this is the timescale most relevant for policy makers.

IB: What other legal frameworks govern deep-sea corals?

MR: All the hard corals that build reefs, they're all listed under Appendix II of CITES—it's a blanket listing, actually. And it's a bit of a pain, in my mind, because the gorgonian [i.e. soft] corals—which include precious corals like *Corallium* that we have been polishing into jewelry since antiquity— are not listed under CITES, although they're the species people are trading and [that] are in great trouble. So that's another political hot potato. The irony is that some of the listed corals—like *Lophelia*—are a bit like weeds in the ocean, they're growing everywhere. But if I wanted to send that to you to look at, I'd have to arrange an export permit, and you would then need to get your import permit [through CITES]. It sounds simple, but the reality of it is that it stops the science dead in its tracks.

IB: What is your position about collaborating with oil and gas companies?

MR: My lab does quite a lot of work with oil companies. We look at the datasets that they generate—they have to do those because of the laws here and in other parts of the world, and because they want to show corporate responsibility. Norway is the richest country in Europe by a long shot because of the petrol dollars that they earn. One of the reasons we know so much about Norwegian deep-water corals is because their oil companies mapped them when they put in their pipelines. They've had to put complex drilling operations together and they've had to bring pipelines up that then snake around the deep-water coral reefs.

This [underlying economic interest] is true all the way back in time to the first deep-sea expeditions. Why did these expeditions get going in the first place? Well, the global superpower at the time was the United Kingdom, [which] put the navy onto it. And why did they bother to do it? Well, they were thinking about transatlantic telegraph cables. They were wondering why the cables broke, because they didn't know about the ridge in the mid-Atlantic and the complex topography. The good thing was that they fused those industrial needs with academic curiosity. The major question they asked was "Is there life in the deep sea?" So, in 1870, the *Challenger* went to sea and deep-sea corals were found on that expedition and were written up in the reports. But they're a kind of Cinderella species: we've been aware of them for quite a long time, but it has taken the technology and the maps and the capacity we've had only in the last two or three decades, really, to reveal just how extensive they are. Still, there are very few of us studying this unique ecosystem. When the deep-sea coral community assembled at the international meeting in 2016, there were probably two hundred people, which is nothing.

CHAPTER 5

The Coral Holobiont

Hope and the Genomic Turn

Without collaborations, we all die.

—Anna L. Tsing, *The Mushroom at the End of the World*[1]

SYMBIOTIC BEGINNINGS

The relationship between coral and algae, which began some fifty million years ago, is one of life's classic symbiotic systems. First coined by Heinrich Anton de Bary in 1879, the term *symbiosis* articulates "the living together of different species."[2] The symbiotic system usually comprises a large partner, the *host*, and smaller partners called *symbionts*. As some have pointed out, this size-based distinction is quite arbitrary: in terms of both cell numbers and genome size, the microorganisms outnumber their host.[3] In the case of corals and sponges, for example, the ratio of host genes to symbiont genes is one to four.[4] Symbiosis can take many forms and shapes. Typically, however, it is character-ized by different levels of mutualism, in which both the host and the symbiont benefit from the interaction.

The symbiotic relationship has radically modified the life of scleractinian (or stony) corals, who belong to the cnidarians—a large group of invertebrates that sting to catch prey and can take either a polyp or a medusa form. Spe-cifically, it has allowed these coral organisms to develop in shallow waters with low levels of nutrients, resulting in the largest bioconstruction on earth: coral

---

FIGURE 35. Hybrid coral *Acropora prolifera* grows in a shipping channel called Government Cut at PortMiami, Florida, August 2011. Although neither parent species expresses green fluorescent proteins in the wild, the hybrid demonstrates a high degree of these proteins. Courtesy of Coral Morphologic.

reefs. Almost half of cnidarians harbor photosynthetic symbionts both in their cells and in the extracellular spaces within their tissues. These may be acquired either by vertical transmission (from parents to progeny) or horizontally (through the environment).[5] While symbiosis conveys important benefits, it also requires costly adaptations: the host must live in an environment that is highly exposed to light, it must absorb CO_2, and it must reject oxygen.[6]

When I asked Ruth Gates, director of the Hawai'i Institute of Marine Biology and president of the International Society for Reef Studies, why she decided to study corals rather than any other taxa, she explained that corals encompass "all of the biological complexity you find across the entire animal tree." "They do everything," she said, pointing to how coral hosts use other organisms' biology to their advantage. She was particularly impressed by the corals' level of evolutionary innovation. "A plant living inside an animal cell? Come on, that's science fiction! The plant lives there, but it feeds the animal from within. Talk about cooperation! We [humans] can't do that. Think about it, if we had plants living in our bodies, we'd solve the food problem. Just imagine it!" Gates's enthusiasm was contagious. She went on to tell me about the ability of corals to change shape in response to their environment, which she referred to as *morphological plasticity*. "We all have hard parts all right," she explained. "But we don't change the shape of our hard parts often. Instead, our soft spots change. But corals can do all of it. They can become willowy branches or, if conditions like flow rate increase, they hunker on down or become short stubby little branches in the same individual. That's crazy stuff right there."[7]

The "crazy stuff" associated with corals has become more visible to humans with the advent of genetic and genomic tools—in other words, tools that apply to the study of both individual hereditary traits and the entire organism's genome. These new technologies have created challenges to the traditional classification of species, and even to Darwinian ways of thinking about evolution. Advances in microbiology have also allowed scientists to better explore the symbiotic relationship within stony corals and the role of symbionts in coral resilience. The third global bleaching event of 2014–2017 has lent these studies greater urgency, as some scientists argue that they may have practical applications for enabling coral survival into the next century. What we are witnessing, arguably, is the genomic turn in coral conservation.[8]

SYMBIONTS AND HOLOBIONTS

After speaking to coral scientists for several years, it has become apparent to me that animals and plants—as well as all other living organisms—can no longer be considered individuals, at least not neatly so and without considerable back-bending. These organisms are, instead, symbiotic conglomerates, or holobionts, with independent evolutionary significance. This insight, most famously advanced by microbiologist Lynn Margulis, has challenged the grounds of individual-based biology and the political science that has emerged alongside this biology.[9] It has also very much molded the field of coral science.

Margulis demonstrated that microbes are implicated through symbiosis and cell fusion in the emergence of biotic variety at all scales in a process named *symbiogenesis*. In 1991, she proposed that physical association between individuals of different species for significant portions of their life history is a "symbiosis," that participants in the symbiosis are "bionts," and that the resulting assemblage is a "holobiont."[10] In their 1995 book *What Is Life?*, Margulis and her son, Dorion Sagan, offered that "the strength of symbiosis as an evolutionary force undermines the prevalent notion of individuality as something fixed, something secure and sacred. A human being in particular is not singular, but a composite. Each of us provides a fine environment for bacteria, fungi, roundworms, mites, and others that live in and on us."[11] This symbiotic reality implies a new understanding of the natural world in which symbiogenesis plays a foundational role as an evolutionary mechanism.[12]

These biological realizations soon trickled down to the humanities. In their 2012 article "A Symbiotic View of Life: We Have Never Been Individuals," biologist Scott Gilbert, historian of science Jan Sapp, and philosopher Alfred Tauber drew a correlation between the theory of an autonomous individual in biology and the parallel appearance of the modern independent citizen. They argued that the notion of symbiosis shatters the foundations of neo-Darwinian biology and challenges the accepted essentialist views of individuality that have characterized both biology and Western thought more generally. In their words,

> In the early modern period, mirroring the appearance of the independent citizen, the notion of the autonomous individual agent framed a biology that was organized around the study of particulate, interacting, living entities. . . . These discoveries have profoundly challenged the generally accepted view of "individuals." Symbiosis is becoming a core principle of contemporary biology, and it is replacing an

essentialist conception of "individuality" with a conception congruent with the larger systems approach now pushing the life sciences in diverse directions. These findings lead us into directions that transcend the self/nonself, subject/object dichotomies that have characterized Western thought.[13]

Gilbert and his colleagues further argued that only with the emergence of ecology in the nineteenth century did organic systems, composed of individuals in cooperative and competitive relationships, complement the individual-based conceptions of life sciences. New technologies such as the microscope and, much later, polymerase chain reaction and next-generation sequencing, have revealed a rich world of intermingling relationships among microscopic and macroscopic life. In addition to the systems approach, "the discovery of symbiosis throughout the animal kingdom is fundamentally transforming the classical conception of an insular individuality into one in which interactive relationships among species blur the boundaries of the organism and obscure the notion of essential identity."[14]

The recognition of the centrality of symbiosis and holobionts is part of a substantial paradigm shift in biology, which until recently was dominated by neo-Darwinian theories of origin and natural selection. It is also a move away from the Koch/Pasteur school of microbiology, with its focus on disease, toward the environmental school, which centers on the interactions between a variety of organisms as well as their environment.[15] Finally, the shift toward the holobiont—the conglomerate of a host and its symbiotic organisms, including viruses—also enabled the reemergence of Lamarckian evolution as a challenge to the traditional Darwinian scheme. Although he had no idea about microorganisms, Jean-Baptiste Lamarck—whose evolutionary theory preceded that of Charles Darwin—believed that the environment could lead to changes in organisms that would then be inherited. Current Lamarckian theory considers evolution that is primarily driven by cooperation among and competition with microorganisms, and not so much by mutations. Along these lines, modern evolutionary developmental biology (or evo-devo) questions the idea that evolution is necessarily gradual. This new scientific movement recognizes that much evolution has been "saltational" (that is, capable of making jumps), sudden, and "punctuated."[16]

THE CORAL INDIVIDUAL

Who is the individual coral? Among the coral scientists I interviewed for this book, the responses were considerably varied. Although she recognized that "the concept of an individual in the coral gets really complicated," coral geneticist Iliana Baums nonetheless contended that "the exact definition of the individual would be the polyp."[17] But from an evolutionary perspective, the individual is insignificant, she qualified, proposing instead that the genotype is the relevant evolutionary unit. In her words:

> If you look at a colony, it's made up of many different individuals, but they're all genetically identical because they're produced through budding and asexual processes. The colony itself, the fragments, grow into a new colony, which is genetically identical as well and so it is, again, a clone. The only time you get a new genotype is when you take the eggs and sperm of one colony and cross it with something that is genetically different. Across the Caribbean, if you were to sample one thousand colonies, you would recover about five hundred genotypes. But then Florida's many reefs are made up of only one clone. Evolutionarily speaking, those reefs are dead unless a new genotype arrives and settles there.[18]

For these reasons, Baums tries to avoid using the term *coral individual* in scientific publications. Instead, she told me, she prefers to borrow the terms *genet* and *ramet* from the plant literature. She explained that

> *genet* counts the number of genetically different genotypes in this population [and] *ramet* is used for the modules [or units] that share that same genotype. For example, you can have a reef that has, in plant terms, one hundred ramets on it, [which is] just one genet. *Genet* and *ramet* work well as terms; they're just not something that a lot of people are used to for an animal. It's very challenging to talk to the public about it, because people don't care much about plants.[19]

Like Baums, Mikhail Matz is also a coral geneticist; but unlike Baums, he identifies the coral individual at the level of the colony and not at that of the polyp. From his point of view, the polyp resembles a human finger, except in the coral's body. In his words,

> Corals are just like any other organisms, just like your cat and dog. It's an individual. One coral colony is an individual. Polyps are just like my fingers, they don't count. So one colony is one individual. Very rarely does it happen that a branch falls off and grows into another colony. But that's not how the corals reproduce normally. Normally, they do one big orgy: once a year they spew the gametes into the water column, all together. The only thing which needs to be taken care of is synchrony, because you need to do it together with all your neighbors. So they do this once a year, synchronously, and gametes recombine, just like in any other animal. Corals are actually pretty good animals as far as model systems go. And after fertilization, what results is tiny little larvae that float about with the ocean currents for a couple of months. Then they settle

on the reef and the tiny coral larvae grow into individual colonies. Each colony comes out as a result of a larva growing, growing, growing into a larger and larger colony. But it's the same genetic individual that is the result of one single sexual reproduction.[20]

I shared Matz's comparison of polyps to human fingers with John Finnerty, Les Kaufman, and several members of their labs at Boston University. They were not convinced. "The finger metaphor is not helpful," Finnerty explained. "The cells in that finger, their only mechanism for propagating that DNA is as part of that unit. But the polyp can live on its own and has the capacity of forming a new unit. Polyps are also interchangeable: fingers can't do what toes can do. But a polyp is a polyp is a polyp"[21] (see, e.g., figure 36). Although he basically agreed with Finnerty's statement, Kaufman qualified that in certain species, differentiated polyps are what enable corals to grow into certain reef structures, such as branches in the case of *Acropora cervicornis* and blades in the case of *Acropora palmata*. The discussion lingered on for a couple of hours.

As we sat together in the small conference room, five coral scientists and myself, on a nice April spring day in Boston, I couldn't help but feel jealous of both the intellectual rigor as well as the strong sense of commitment, engagement, and belonging that these scientists share—their interconnecting tissue, to borrow from the coral realm. Are they more like fingers or more like polyps, I mused.

The question of coral individuality is not a new one. It was already contemplated by none other than Karl Marx, who wrote in his 1867 book *Capital* (albeit in a footnote): "In corals, each individual is, in fact, the stomach of the whole group; but it supplies the group with nourishment, instead of, like the Roman patrician, withdrawing it."[22] Elsewhere in the book, Marx wrote: "We see mighty coral reefs rising from the depth of the ocean into islands and firm land, yet each individual depositor is puny, weak, and contemptible."[23] Who would have guessed that thinking with corals would lend support to Marx's ideas on the power of the collective?

The debate around the definition of the coral individual is not only conceptually interesting, but also holds physical, normative, and political significance. As I have shown in chapter 4, to ascertain if they are threatened or endangered under environmental law, individual corals must be both identifiable and quantifiable.

FIGURE 36. *Pocillopora damicornis* (cauliflower coral) polyps captured with a Zeiss laser scanning confocal microscope at 2.5× magnification. The darker parts (red in original) are the single-celled *Symbiodinium* living inside the coral polyps. The lighter parts (green in original) are green fluorescent proteins, seen throughout the coral tissue but most prominent in the coenosarc—the tissue between the polyps. At the tips of the tentacles and around the mouth in the center are nematocyst (stinging) cells (blue in original). Photo by Amy Eggers. Courtesy of Gates Coral Lab.

THE CORAL HOLOGENOME

The symbiotic relationship at the core of their existence makes corals model organisms for the study of symbiosis. It's no coincidence, then, that the proponents of the hologenome concept in biology were coral scientists. Eugene Rosenberg and Ilana Zilber-Rosenberg researched mechanisms of coral infection for years, and in 1996 they observed corals becoming resistant to certain infections and also to bleaching. A decade or so later, they published the "coral probiotic hypothesis," which suggests that "a dynamic relationship exists between symbiotic microorganisms and environmental conditions which brings about the selection of the most advantageous coral holobiont."[24] They then coined the term *hologenome*, which posits that "the holobiont (host + symbionts) with its hologenome (host genome + microbiome), acting in consortium, function as a unique biological entity and therefore also as a level of selection in evolution."[25] In 2003, Nancy Knowlton and Forest Rohwer adopted the term *holobiont* to describe the relationship between viruses, bacteria, archaea, dinoflagellates, and corals.[26] The acquisition of heat-resistant *Symbiodinium* in the coral fits within this concept and explains how coral microorganisms can assist with adaptation, possibly accounting for the evolutionary success story of corals.

Christian Voolstra is a German microbiologist and ecologist working at the KAUST Institute in Saudi Arabia. He is interested in the intersection of coral science and microbiology. As he explained in our interview,

> The coral is a prime example of symbiosis between a higher-order animal and intracellular, photosynthetic algae. It's one of the few examples we are aware of where somebody lives in your cell and it's mutualistic, not parasitic or pathogenic. Somehow, corals harvest this relationship to their benefit, or, at least, it's a give-and-take. Now, the bacteria people are coming in and saying—everything is symbiosis. The coral people [already] have fifty years of research into symbiotic relationship. If you put [the coral scientists] together with [the] microbial people, you can make big strides in understanding this because corals are even better than other systems in that they have bacteria—the microbes—but they also have intracellular, nucleonic symbionts.

"If you want to learn about holobionts or metaorganisms," Voolstra concluded, "corals are the prime example."

The term *metaorganism* and its close relative, the *superorganism*, highlight that single-species groups, multispecies communities, and, by extension, human societies can possess the properties of single organisms.[27] Within biology, social insect colonies have been regarded this way for centuries.[28] More recently, some have compared the succession of plant species that culminates in a forest to the growth and development of a single organism. James Lovelock took this idea even further, portraying the entire earth as an organism who regulates her atmosphere to be conducive for life—what has come to be known as the "Gaia hypothesis."[29]

Corals have been central to studies of the metaorganism, as these early forms of life only create an intact ecosystem when in harmony with their microbial populations.[30] In addition to studying the life forms embedded within the coral, coral scientists often point to the unique forms of life in the ocean environment. Unlike terrestrial species, corals literally "swim in a sea of microbes," Voolstra explained. At the same time, many of them lack mobility, being sessile organisms. Margaret Miller of NOAA thus emphasized the importance of the ocean environment for understanding corals. Her focus, however, has been on the centrality of health and disease. "[Humans] have skin and tissue, and all of our basic physiological processes are well protected within a homeostatic environment," she explained.[31] By contrast, many corals live with mucus layers on their outside surface, and so the interactions between the animals, their symbionts, and the environment are much more direct. The corals are "exposed to all the bacteria in the ocean, exposed to all the toxins, [and to] all of the environmental variability that the ocean water contains." Finally, Miller pointed out that in spite of the critical need to better understand coral-related diseases, "we've made very little progress in understanding what causes coral disease" and, in fact, "we're sort-of in the medieval dark ages [in this regard, like] when we were using leeches to heal human disease."

Despite the growing importance of microbial and genomic research (a lot of which indeed focuses on disease) for coral science, many coral geneticists still complain about what they feel is a bias by the majority of coral scientists against genetic work in the lab, and a strong preference toward practicing marine biology in the field. As one interviewee anonymously told me,

> The coral field is doing a lot of lateral science. They describe how horrible it is and how the reefs are going down. But, effectively, they are not trying to use all the tools available. There's resistance to studying and understanding the whole [coral] thing, and a strong romantic association with going into the field. There is tremendous value [in] working in the field and seeing how organisms behave in their environment; it gives you a better perspective. [However,] I think a lot could be better understood if you spent the money to do [genetic] research in the lab. The study of the coral as a holobiont is not this romantic thing that is coral reef research in the ocean. It just pisses people off when you say it to their face. The biggest coral reef conference, held every four years, will be held in June [2016] in Hawai'i. There are four thousand scientists. If you compare the molecular sessions to the traditional sessions, it's a ratio of one to one hundred.[32]

Michel Pichon, one of the founders and elders of coral taxonomy, agrees about the rising importance of genetics in coral research, emphasizing in particular the invaluable application of genetic tools for understanding coral speciation.[33] In his words,

> The first people to be interested in coral taxonomy were geologists and paleon-tologists. Way back in the nineteenth century, they laid down the framework, which has since then been refined, of course. Even after World War II, [the] scheme of classification [was] based on both the original shape of the colony and the final structure of the individual colonies. Because it was, in some ways, so easy [to pay attention to the structure], nobody really paid much attention to the living parts. From a taxonomic standpoint, the living part of the coral had pretty much been ignored until the [emergence of the] genetics school about twenty years ago.[34]

The turn of coral science toward genetics and genomics not only informs existing coral classifications but, as we will see, also raises critical questions about their continued relevance.

CORAL HYBRIDS AND RETICULATE EVOLUTION

John Veron identified and named more than twenty percent of reef corals worldwide. He recently completed a massive online database project, Corals of the World—a dynamic account of his three-volume book of the same name.[35] Veron earned his nickname, "Charlie," for being perceived as the Charles Darwin of the coral world.[36] In our interview, however, he challenged Darwinian theory as it applies to coral speciation, instead emphasizing neo-Lamarckian ideas of reticulate evolution in marine organisms. In his words,

> Corals don't actually fit into nice little units. They vary geographically, and they interbreed here and they evolve there, so they form these interlocking patterns in time and space. And that means that if you do a very detailed taxonomy here, the taxonomy over there is more or less the same, but not quite. I had worked out all of eastern Australia and then went to western Australia and sat questioning my original work. Things I had said weren't holding up. Some species were the same but they didn't have the same relationship with neighboring species. I couldn't separate them, or they seemed to be very different. And so I got to the stage where I was ready to chuck it in. Scientists like to be right. And I was never right. I was always mostly right, sort of, more or less.[37]

His frustration with what appeared to be his inability to correctly classify corals into a consistent taxonomy led Veron to an important eureka moment that has redefined his entire career: "One morning, I was at work very early, half asleep, boiling water, and the whole concept of reticulate evolution came into my head, fully formed, in every ounce of detail, exactly as it is now. Basically, it was blindingly obvious, but it hadn't occurred to me and it hadn't occurred to anyone else." Within the paradigm of reticulate evolution, he explained,

> species have fuzzy boundaries—geographic, morphologic, and genetic—and they have fuzzy interactions with other species. And it's the very opposite of Darwinian evolution. Darwinian evolution is obvious. [But] currents move genes around, they move larvae around. So the genetic pathways, the pathways of larvae, are the pathways of current. And currents change all the time. So the template is changing for all species, except those that can swim independently from currents. . . . The units that are formed, or the species, if you like, are not the same as [those that] formed the last time; they reshuffled the pack. It's the same genes but it's all mixed up. It is a reticulate pattern. You can call it hybridization, but it's also true that every species is a hybrid, everything is a hybrid. So [in] Darwinian evolution you have a species forming from a point of origin in time and space. But [in] reticulate evolution there is no origin point: there are genes moving constantly, all the time.

Veron's embrace of the prominence and importance of coral hybrids, and of reticulate evolution more broadly, is not shared by the entire community of coral scientists. The story of one coral species, *Acropora prolifera*, attests to the debates within this community about the nature of coral evolution. Although highly theoretical, these debates also inform the degrees and types of interventions that are deployed on the ground—or, rather, underwater—to save corals.

ACROPORA PROLIFERA

The Caribbean *Acropora prolifera* is the only coral hybrid in the world that is widely recognized as a species. It is a hybrid of the two threatened species, *Acropora cervicornis* and *Acropora palmata*.[38] Nicole Fogarty of Nova Southeastern University in Florida is the prime expert on this coral. Referring to herself as Lady Hybrid, Fogarty readily admits to her emotional ties to the *prolifera* corals. "I feel like they're my babies," she told me in our interview.

> No one else in the world knows them better than I do. I can't tell you how many days on end I've stared at these hybrids, my legs soaked in the shallow waters where they grow. We've developed a very special bond. They are beautiful corals and are doing phenomenally well, especially considering the poor state of their parent species. We've been so focused on trying to restore those threatened *Acropora*

species, that a lot of people totally missed out that they're being restored naturally by the hybrids. There are a lot of skeptics out there. These hybrid corals are too rare, they say. But these skeptics are not [out] in the field; they haven't really seen the healthy patches of *prolifera* in the Caribbean. And [I predict that] we are going to see more and more of them as the conditions in the water change [see figures 37 and 38].[39]

Further into my conversation with Fogarty, more details about her col-leagues' skepticism toward *Acropora prolifera* emerged. As it turns out, this hybrid has been the subject of a major debate between two camps: Steven Vollmer and Stephen Palumbi, on the one hand; and Madeleine van Oppen, David Miller, and their colleagues, on the other hand. In their 2002 article in *Science*, "Hybridization and the Evolution of Reef Coral Diversity," Vollmer and Palumbi argued that DNA sequence data from the three *Acropora* corals growing in the Caribbean in fact show that mass spawning—whereby differ-ent *Acropora* species synchronize their gamete release—does not erode species barriers. Specifically, these authors posited that *Acropora prolifera* "are entirely F1 [i.e., first-generation] hybrids of these two species, showing morphologies that depend on which species provides the egg for hybridization." At the same time, they acknowledged that "F1 individuals can reproduce asexually and form long-lived, potentially immortal hybrids with unique morphologies."[40] Referring to the *prolifera* hybrid as an "immortal mule," Vollmer and Palumbi portrayed this coral as an evolutionary dead end from a Darwinian viewpoint. Accordingly, they proposed that hybridization has little evolutionary potential.[41]

FIGURE 37. Hybrid coral *Acropora prolifera* at Carrie Bow Caye Island, Belize. Nicole Foga-rty, of Nova Southeastern University, reflected on this image: "This is a one-acre island leased by the Smithsonian. Scientists have been going there for decades to do marine research. I've been going there for twelve years and it is near and dear to my heart. This field station is a unique research study because it is very remote and off-the-grid—only six scien-tists can visit this island at a time. Belize is not immune to the stresses that impact the rest of the Caribbean: disease and storms and overfishing and coral bleaching. Yet, Belize's coral reefs are some of the most beautiful that I have seen in the Caribbean. I have studied acropo-rid corals off Carrie Bow since 2005, spending countless hours scuba diving and snorkeling there—so I have gotten to know the corals pretty well. It is very hard work, physically. I have never worked harder in my life. We night-dive for the coral spawning, then bring the sperm and egg back to the lab to conduct experiments all night, then the next day we continue our experimentation and research, and then we are up again the next night. The sleep deprivation is intense. As you can see in the photo, the hybrids are right off the shore, so all I need to do is throw my gear on and my study organism is right there. I like this image because it conveys how shallow and therefore how robust these corals are—the shallows cool off in the winter and are hotter in the summer, therefore the temperatures are more extreme. Another thing I like about this photo is that you can see the island in the background. Throughout the Carib-bean, these hybrid corals live close to human development and they thrive nonetheless. Thinking about global climate change and the future of our reefs, these robust hybrid corals may help us understand the corals of the future. These young corals, perhaps the youngest coral taxon in the world today, have the potential to persist beyond the parents. This gives me hope" (via Skype, June 15, 2017). Photo by Abby Wood, c. 2009, Belize.

Madeleine van Oppen and her colleagues disagreed. While they acknowledged that the *prolifera* coral "backcrosses with the parental species at low frequency,"[42] they nonetheless suggested that "although the effects of reticulation are ... relatively small when measured on ecological time scales, on evolutionary time scales these rare hybridization events are significant."[43] They were especially unhappy with the "immortal mule" analogy advanced by their colleagues. Their response:

> The analogy of "evolutionary mules" is also inappropriate and unhelpful in the context of coral evolution. Many lineages of organisms are recognized as distinct species yet (unlike mules) reproduce asexually, and some of these distinct species are the products of hybridization. Species that are distinct but reproduce asexually evolve in the sense that they accumulate mutations; they also compete for resources and may persist over evolutionary time. Clearly such asexual lineages can be important for the evolutionary process, and should not be dismissed as irrelevant simply because they do not fit with the usual view of how species should reproduce and evolve. Rather than mules, parthenogenetic lizards [i.e., lizards who reproduce asexually by recombining their own DNA material] are a more appropriate vertebrate analogy for *A. prolifera*; for example, the lizard genus *Lacerta* contains seven diploid parthenogenetic species of hybrid origin.[44]

The debate about whether the *prolifera* hybrid is more mule or more lizard, while seemingly pedantic, in fact carries with it considerable practical implications. Traditionally, mainstream conservation approaches have tended to see hybrids as dead ends and thus as undesirable from an evolutionary standpoint—a tendency that has resulted, for example, in an official policy by the U.S. Fish and Wildlife Service to avoid hybridization in captive breeding efforts carried out under the Endangered Species Act. In 1987, this policy led to the infamous extinction of the dusky seaside sparrow (*Ammodramus maritimus nigrescens*).[45] Almost twenty years later, the agency in charge of implementing the Endangered Species Act in the marine environment, NOAA, declined a petition to include the *prolifera* coral in the statutory protection list because, as a hybrid, it failed to meet the act's definition of a species.[46] The same underlying policy advanced by NOAA has resulted in the agency's refusal to grant permits to Florida coral nurseries to propagate *prolifera* for purposes of transplantation onto wild reefs.[47] In a 2006 paper, van Oppen and her colleagues called for further studies of the role of hybridization that would better inform conservation management strategies. In their words, "We conclude that outcomes of hybridization are significant for the future resilience of reef

FIGURE 38. The hybrid *Acropora prolifera* can be seen up front, and the dead parent species *Acropora cervicornis* and *Acropora palmata* are in the background. Nicole Fogarty told me in our interview: "This signifies the dramatic shift that we are seeing: after the demise of most of the elkhorn and staghorn, we are witnessing the increase of this more robust hybrid at some sites. The hybrids seem to be more tolerant of heat and disease. As much as I would love to have [the hybrid] protected, I understand that this coral is everything but endangered. In fact, it is the opposite: this, potentially, is the future of shallow coral reefs" (via Skype, June 15, 2017). Courtesy of Nicole Fogarty.

corals and warrant inclusion in conservation strategies."[48] In our interview, van Oppen outlined the broader context of her position on hybrids:

> In traditional conservation, hybrids are not even considered because they're not pure. I don't think we can afford this purist view. It's a closely related coral, it's functional in the system, and it's coping with these disturbances—that's a better solution than having no coral. I think we will, out of necessity, move away from the purist view.[49]

Her current research reflects this position. At the Australian Institute of Marine Science's National Sea Simulator (SeaSim), van Oppen has been breeding hybrids from eight parent *Acropora* species, then spawning and fertilizing them in an attempt to create first- and even second-generation offspring. She believes that "studies of animals with plant-like life histories will enhance current understanding of the evolutionary significance of hybridization in animals." Indeed, "hybridization is by no means as rare or as dead-end or as dependent on asexual reproduction as previously thought," evolutionary biologist Stuart Newman explained.[50] For example, he told me, many bird species[51] and primates[52] "are the fertile (sexually reproducing) outcomes of hybridization events."

Alongside hybridization, chimerism can also account for alterations at the coral's genomic level, and thus presents serious challenges to classification. A chimera is an organism containing tissues or cells of at least two genetically distinct conspecifics (members of the same species).[53] An emerging field in contemporary coral science, the extent of chimerism in wild populations of reef corals is still unknown. Nonetheless, some tend to see chimerism everywhere. "The entire landscape of life—from cells to biomes—is substantially an evolving collection of chimeric opportunities," science educator Douglas Zook wrote.[54] A recent study estimated chimerism at about 5 percent in natural populations of *Acropora millepora* from Australia, suggesting that chimerism is more widespread in corals than scientists previously realized.[55] In our interview, Ruth Gates told me about the confusing results that she received from a genetic lab with regard to her coral samples:

> We [wanted] to find out the genome of this individual. [So] we took the normal, very high-efficacy, approach of collecting all the sperm from one colony and said, "Okay, this is one individual," and sent it off [to the lab]. When we got the data back, it was so confusing [that] we couldn't figure out what the hell was going on. We went back and collected individual sperm bundles from individual polyps across the same colony and we found they were different. That's wild.[56]

The latest results from the lab, Gates suggested, indicate that a single colony of certain coral species may actually be chimeric—comprising two individuals, and perhaps more. Chimerism has recently been recorded in humans and may be happening more often than we realize,[57] challenging our most fundamental sociopolitical and legal assumptions about the centrality, and indeed the very meaning, of the individual.

ASSISTED EVOLUTION AND SUPER CORALS

The term *assisted evolution* was first coined in the general conservation literature in 2009 and defined there as "the acceleration of natural evolutionary processes

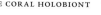

to enhance certain traits."[58] Madeleine van Oppen and Ruth Gates have been its leading proponents in the coral context. For them, *assisted evolution* is a holistic term that incorporates genetic, epigenetic, and microbiome evolutionary changes.[59] Here is how van Oppen explained this term in our interview:

> The full term is *human-assisted evolution*. You're basically assisting natural evolution, trying to push it a bit. It captures the essence of what we're trying to do very well. We are tackling a range of processes that occur in nature. Assisted evolution captures all of that—selective breeding, manipulating microbial communities—it captures all of these processes and approaches.[60]

Gates broke it down further for me. First, she assured me that there is nothing unnatural here, and nothing that hasn't already been done before. "Really, the basis of our project is [that] we're assisting nature, we're accelerating natural processes. We're not creating anything new. We're doing what nature does, we're just trying to find ways to do it more quickly. . . . [And] we're doing that through selective breeding, just classic selective breeding." Gates also told me about the process of selecting for certain traits that could withstand future conditions. "We're interacting very hardy genotypes," she said. "We're doing that through conditioning: running our corals on environmental treadmills, with the idea that what doesn't kill you makes you stronger."[61]

In their 2015 article, "Building Coral Reef Resilience through Assisted Evolution," van Oppen and her colleagues proposed four approaches to developing coral stock with enhanced environmental stress resistance: random mutations, natural selection, acclimatization, and changes in microbial symbiont communities. In a follow-up publication from 2017, van Oppen explained that "it is important that assisted evolution becomes embedded within coral reef restoration initiatives, because the worldwide extensive loss of coral cover suggests natural rates of evolution of stress tolerance are too slow to maintain functional coral reef ecosystems into a future characterized by rapid climate change."[62]

At the same time, van Oppen, Gates, and their collaborators have assured their colleagues that "as with any restoration initiative, assisted evolution approaches need to be guided by historical information, contribute to the restoration of ecological structure and function, and . . . have the ability to adapt further to contemporary selection pressures."[63] Furthermore, they emphasized that "our guidelines follow those of rangeland restoration practitioners, who recommend that in the development of more resilient stock, the most 'local' options must always be considered before any nonnative ones. . . .

We suggest a process that considers the lowest levels of intervention first and progressing to more aggressive intervention only when necessary." More than anything else, such extensive assurances indicate the trepidation of the broader coral community toward such interventions when performed at the cellular level.

Still at the cellular level, the field of epigenetics is currently emerging as the cutting edge of coral research and restoration efforts. Simply put, epigenetics is the study of changes in organisms caused by the modification of gene expression, rather than by changes in the genes themselves. Van Oppen explained in our interview that epigenetic changes usually happen "a little bit quicker than truly genetic revolution." Furthermore, she told me that "there's evidence from a range of organisms that you can induce genetic changes and that some can be passed on from one generation to the next." These realizations have led van Oppen to ask: Can we precondition adult coral colonies by exposing them to predicted future ocean conditions? Would that make their offspring more tolerant to these conditions? She designed a large experiment to find out (figure 39).

Van Oppen's experiment is set up to study nine systems in three time periods: current, mid-century, and late-century. Using predictions by the Intergovernmental Public Panel on Climate Change for ocean chemistry conditions in 2050 and 2100, the experiment will track all individual offspring of a single species of brooding coral, *Pocillopora acuta*. Van Oppen told me in our interview that "the advantage of having the species that produces asexual broods is that all the offspring of a single colony will be genetically identical. So we can distribute those to all of the nine systems that we have in the end and that will help us tease out the genetic versus the epigenetic effects."[64]

Another experiment designed by Gates and van Oppen involves "playing" with various coral components to effectively create "super corals." As straightforward as it may sound, the term *super coral*, and its recent deployment in the context of assisted evolution and by the popular media, is already contested and thus necessitates a brief explanation. According to Gates, "Hybrid vigor has been known to science for decades. There has been a lot of science about the variation in performance in individual coral species and in hybrids [as well as] repeated observations of extreme performers from places like the Red Sea and in power plant effluents in Taiwan."[65] However, Gates argues that the use of this term in the context of assisted evolution (or AE, as she calls it) is slightly different. In her words,

FIGURE 39. Up front are a variety of propagated acroporid corals prepared for experimentation at the National Sea Simulator in Townsville, Australia. In the background are the Transgenerational Change experimental systems, which focus on a single species of brooding coral, *Pocillopora acuta*. The scientists control temperature (daily and seasonal) and partial pressure of carbon dioxide (daily and seasonal cycles). The lighting systems for the Transgenerational Change experiments replicate daily, seasonal, and lunar cycles, triggering the corals to release larvae in a predictable manner. Photo by Christian Miller, May 2016. Courtesy of the Australian Institute of Marine Science.

> We rapidly increase the performance of talented athletes so they become super athletes by identifying them, training them in the gym, giving them great nutrition and then they often breed with people they meet in their training environment to produce extraordinarily talented offspring. This is exactly what we are doing with the AE work: identifying the best of the best [this is the older meaning of "super corals"] and then bringing them into the lab to train them on environmental treadmills (intragenerational acclimatizations), providing the best nutrition (manipulations of the zooxanthellae and microbes), and finally selectively breeding them to genetically direct very high-performing offspring (transgenerational acclimatization).[66]

In the vocabulary used by proponents of assisted evolution, super corals are superior not only because they are naturally sturdier but also because they are then engineered to bolster that natural sturdiness. "Super corals are AE engineered corals," as Gates put it. Past research had revealed that extreme performers exist in nature, and the assisted evolution project builds on this vigor to create further vigor. In scientific terms, Gates's super coral is an F2 (second-generation) coral.[67]

One of the ways assisted evolution is utilized to create a second-generation super coral is in the making of super algae. In the words of van Oppen,

> I have become more and more of a microbiologist over the years, because microbes are so important. We're trying to evolve microbes. At the moment, we're only doing it with the algae. We can isolate the algae from the coral, culture them [and] subject them to selection. For example, we grow them at higher temperature over many, many generations. Every time a cell replicates, the genome is copied by enzymes and those enzymes make mistakes. Not all these mistakes are repairs, you get somatic mutations, too. Given enough mutations, you may have a beneficial mutation, for instance in terms of how this organism performs under higher temperature. Those cells will then outcompete cells that haven't mutated. You grow that out. We've been playing with that. We have some exciting preliminary results that suggest [that] it might be possible, within a reasonable number of cell generations [and] in some instances, to successfully create more tolerant algal symbionts. We will inoculate corals with that and then test if the coral animal, the coral holobiont, has become more tolerant.[68]

Alongside the scientific and pragmatic questions about their experiments from the largely skeptical community of coral scientists, the central opposition to their assisted evolution project, as Gates and van Oppen see it, is based on the perception that it resembles genetic modification. Highly aware of the widespread distrust toward genetically modified organisms (GMOs), they emphasized again that, at this stage at least, they are merely doing old-fashioned selective breeding.

Still, they admitted that they are open to the possibility of conducting genetic engineering in the future. They like to cite the case of the American chestnut, where "in an alternate approach to the crossbreeding strategy to develop blight-resistant American chestnut trees, an oxalate oxidase gene from wheat was inserted into the American chestnut genome through genetic transformation"[69]—a strategy that proved both efficient and successful. Despite their recognition that genetic engineering techniques tend to elicit considerable public resistance, van Oppen and her colleagues propose that "such approaches may produce desirable results faster and at a lower cost compared to selective breeding." Then again, while admitting that they would be willing to consider GMO corals in the future, they concede that it is not yet time. In their words, "although the development of GMO corals might be contemplated in extremis at a future time, we advocate less drastic approaches that use widely accepted techniques to accelerate naturally occurring evolutionary processes (e.g., random mutations, natural selection, acclimatization, and changes in microbial symbiont communities)."[70] Despite the grave state of reef-building corals, even the most radical interventionists within the community of coral scientists are not developing GMO corals, nor are they using genome editing technologies that have become ubiquitous in synthetic biology, such as CRISPR-Cas 9 or gene drives—not yet.[71]

NITTY-GRITTY GENOMICS AND ASSISTED GENE FLOW

In keeping with recent insights from the field of reticulate evolution, geneticist Mikhail Matz told me about the recent changes in how coral scientists think about time:

> Initially, maybe even five years ago, people really thought that "oh, evolution is so slow, it cannot possibly happen within a few decades." [But] it can happen within a year. It depends on how much variation there is. It's like a grocery store with all sorts of cereal. [When] more hipsters move into your neighborhood, you don't need to gut the whole grocery store and get the new hipster cereal. You already have it. You just increase the order for that particular hipstery thing. That's what's happening: natural selection [is like] a grocery storekeeper.[72]

Despite this grocery analogy, Matz doesn't believe that humans should be assisting the selection process. "I'm so skeptical about this, I'm almost angry," he told me about assisted evolution.[73] "The only way to help things evolve is to ensure genetic variation in the wild. You cannot possibly evolve them on your own, [because] you don't know what to select for! The super coral business is appealing to the public, but it has no biological justification whatsoever."[74]

The most extreme human intervention that Matz would be willing to consider for coral conservation is to spread beneficial mutations across populations by moving them around. "You need to keep it and you need to shape it, but by no means, in no way, can you afford to lose it," he said about genetic diversity. "So instead of moving the gametes, move good potential winners and let them reproduce with other populations to spread the goodness. It's much easier to grab that dude that is resistant to cold and to then let him breed naturally wherever you want to introduce the trait. You don't need to do anything, you don't need any new technologies."[75]

Some scientists have called this type of intervention *assisted gene flow* (AGF), explaining that

AGF, in the form of assisted larval dispersal or assisted adult migration, might facilitate the spread of genotypes with heritable traits—for example, heat tolerance—from warmer to cooler locations or habitats. For example, the northern GBR is approximately 2°C warmer than its southern region and contains genetically heat-adapted corals; thus, northern coral stocks provide the potential to enhance bleaching resistance on southern reefs. An opportunity to climate-harden the northern GBR using AGF could come from warmer seas such as the Persian Gulf. Here, thermal bleaching thresholds of corals are 3–4°C higher than the most bleaching-tolerant assemblages in the Indo-Pacific or the Caribbean.[76]

This form of translocation of existing coral genotypes—or *assisted migration*—has also been referred to by scientists as *genetic rescue*. For example, an article published in *Science* in 2015 stated: "As global warming continues, reef-building corals could avoid local population declines through 'genetic rescue' involving exchange of heat-tolerant genotypes across latitudes."[77] The authors were able to show a tenfold increase in the odds of survival of coral larvae under heat stress when their parents originated from a warmer location.

In his own research, Matz studies how natural evolution works in corals at the genomic level. "If we know where the adaptive variation is, then we can actually put it in models and predict how fast it can evolve," he told me. In addition to providing important information about the current state of corals, genomic analysis can inform scientific knowledge about these coral populations going back one hundred thousand years. "You can actually reconstruct back in time how corals reacted to past climate change. How cool is that?" Matz mused. Whereas existing research typically studies the growth rate of a single individual in a limited life span, Matz told me, he is talking about "the whole coral population of the Great Barrier Reef down to one hundred thousand years ago." Using multiple methods that include a full single genome sequence as well as partially sequenced samples of fifty individuals from each population, he is currently sequencing the Caribbean great star coral (*Montastraea cavernosa*) and staghorn coral (*Acropora cervicornis*).

CORAL AS ARCHIVE

In a related project, Matz has been compiling both genomic and paleontological evidence and cross-referencing the data "to see how we can inform our past understanding of coral populations." He has led expeditions to the Caribbean, where he "digs and sequences" to inform such understandings of the past. "The paleontology collaborators work on an outcrop of the coral fossil reef," he explained. "And I go in and sample the living corals around that same place—and we then sequence and reconstruct their population sizes back in time. We will [finally] see how well these two reconstructions match." Utilized in this manner, coral genomics can provide a window to both the present and the past states of these organisms, while also telling us a lot about their future.

Sandy Tudhope heads the School of Geosciences at the University of Edinburgh. The first time I reached out to him, I mistakenly called him a "tropical coral reef scientist." After that, it took quite a bit of coaxing to get him to talk to me about his work on deep-sea, or cold-water, corals, a vivid reminder of the sometimes tense dynamics between these two communities. Tudhope studies corals as natural archives. Every year, the corals lay down a skeleton of one centimeter or so, he explained in our interview,

> and if you dig a core down through that skeleton, as you go down through the core, you're going back through time. And the skeleton itself has got annual banding in it, a little like tree rings. Usually, it's not visible to the naked eye, so you need to have an alternative way of viewing the coral. We use living corals, and then we try to find old, dead corals so we can get windows, if you like, into previous times.

Between 2010 and 2014, Tudhope found corals in the Galápagos Islands who lived for only one hundred years, approximately five thousand years ago. Studying these corals, he is trying to figure out "what one hundred years of weather looked like five thousand years ago."[78]

From Tudhope's perspective, it doesn't really matter whether the coral is dead or alive; as long as there is a skeleton, he can do the research. "The living ones have tissue outside the skeleton," he explained. "Once that dies, you've got a dead coral colony—but the skeleton is still there." However, life still seems to matter in at least one way: the coral's life span sets a hard limit on the temporal knowledge that one can deduce from her skeleton:

> So if you have individual coral colonies that lived for a century, that's the maximum you can get, and then you have to find an older coral to step further back in time. So, very often, we find that there are gaps between these time tables. We don't have continuous records like with ice, where you can see every year back for ten thousand years. All we have are these windows of a few decades at different periods through time, and we're using those to assess the range of variability.[79]

Still, corals tell stories that ice and trees cannot tell. "With ice, you are limited to either very high latitudes or very high elevation," Tudhope told me. "Trees are really good in mid-latitudes but you can't get much information out of them in the tropics. [And] most sedimentary deposits don't preserve the temporal resolution that you require." Corals, on the other hand,

> tell you about tropical oceans, and the tropical oceans are a major part of the climate system. For example, the El Niño phenomenon is centered around the ocean. So to understand that system, you need to find something that will record that information. You need something that accumulates fast enough that you can see the seasonal cycles [and also] record year-to-year variability.

Performing such "coral readings" through time, Tudhope discovered that the El Niño of 2014–2016 was much stronger than such events have been over most of the past hundred thousand years, including the last big ice age.[80]

Knowing the past is also a springboard for learning how to model and predict the future. "We use the data from the past to test the skills of different types of [climate] models," Tudhope told me. "If they couldn't get it right in the past, they will probably not get it right in the future." In other words, by running different climate models for scenarios that actually occurred in the past, one may gain more confidence in the model's capability to predict future climate. Corals therefore serve as archival objects—their organic and inorganic matter encoding information that enables scientists to acquire knowledge about life and its conditions in the past, present, and future.

CORAL SPAWNING AND CRYOPRESERVATION

Time is crucial for understanding the coral's sexual reproduction, too. About three-quarters of symbiotic corals spawn eggs and sperm for external fertilization in the water column ("broadcast spawners") rather than brood planulae (or larvae) within their bodies after internal fertilization ("brooders").[81] For corals who are broadcast spawners—namely, those who release their eggs and sperm into the water, where fertilization occurs externally (as opposed to brooding corals, who reproduce internally)—the critical moment for fieldwork is the spawning event. Synchronicity is key to the gametes' survival: a difference of one-tenth of a second will determine who will mate with whom, and differences in spawn times, at the scale of minutes, can lead to reproductive isolation among populations or species.[82] Since spawning is the coral's central form of sexual reproduction, if they want to work with coral gametes in the field, scientists must be in the right place at very precise moments.

Sexual reproduction in corals is probably best studied through the *Orbicella annularis* species complex. All three species in this complex—*O. annularis,*

O. *faveolata*, and O. *franksi*—spawn four to eight days after the full moon in late summer. Each colony releases a set of gamete bundles within a few minutes, and most conspecific colonies that spawn on a particular evening release gamete bundles within thirty to sixty minutes of each other. Moreover, the spawning times for these species are remarkably consistent across years and regions. Individual corals spawn at almost precisely the same time each year: the average standard deviation in spawn time for a particular colony is seven minutes for *franksi*, ten minutes for *annularis*, and fourteen minutes for *faveolata*.[83]

How do corals do this without having a clock, not to mention a brain? Coral spawners use a set of cues, Nicole Fogarty explained in our interview.[84] Between 2002 and 2009, she and her colleagues performed a large experiment, tagging and genotyping 488 spawning *Orbicella annularis* over 1,335 observations. These confirmed the existing studies, which established that "patterns of solar irradiance and/or wind fields cue the month, the lunar cycle cues the day, and sunset cues the time of spawning."[85] To this, their work added that "there appear to be random or unknown factors that determine which lunar day a coral spawns (e.g., night 5 or 6), but the timing on any given night is a function of how genetic and local environmental factors interact with the sunset cue."[86] On a more poetic note, the spawning event has been referred to as "one of the marvels of the natural world." The *Huffington Post* observed: "It's like a snowstorm with gravity reversed and the snowflakes are miniature peas" (figure 40).[87]

For coral scientists, the magic is both enhanced and overshadowed by the strenuous and time-sensitive labor required during spawning nights. Linda Penfold of the South-East Zoo Alliance for Reproduction and Conservation (SEZARC) participated in the 2016 spawning event at Ken Nedimyer's nurseries in Key Largo, already discussed in chapter 3 and in his interchapter interview. She recounted:

> At 10 P.M. everything starts to happen and it's just crazy because you've got this really short window of time. And you've got one, maybe two, nights and you've got people out in the ocean, and boats. You've got people rushing in with samples, and this hustle and bustle in the lab. And you're under these enormous sorts of deadlines and pressures, but there's this enormous sense of camaraderie and it's so exciting! . . . You've got literally a few hours for two nights per year to try and figure this stuff out. It's like something out of a fairy tale, really.[88]

For the last several years, Penfold's colleagues from various facilities have flocked to the Key Largo nurseries during the spawning event to conduct

FIGURE 40. *Montipora capitata* (rice coral) spawning in Kaneohe Bay, Hawai'i. Members of the Gates Coral Lab used large nets to capture egg and sperm bundles to perform large-scale selective breeding of the more resilient corals. Photo by Mariana Rocha de Souza, June 2017. Courtesy of Gates Coral Lab.

experiments in larval settlement. In August 2016, Penfold joined them with a particular goal in mind: to increase the levels of coral fertilization and embryo development. Although the bulk of her prior experience has been in assisting rhino and elephant fertilization, Penfold was optimistic. "When it boils down to it," she told me, "a sperm is a sperm is a sperm. They can look vastly different, but the function remains the same, the structure remains conserved, and the end result, which is to get a fertilized egg, is the same, thank god."[89] This is how Penfold explained the experiment as it pertained to corals:

> You separate the sperm and the eggs and then basically you get these big tanks or buckets. And then you mix the sperm and the eggs in the tube until it's a lemonade color, and that fertilizes and makes the embryos. [Next,] you put the embryos into the system and they develop into the larvae. That's how you know it's concentrated enough, but not too concentrated.[90]

Lemonade was not the only food analogy Penfold used; she further described coral gametes as "little mustard seeds" and explained that the early embryo in coral "looks like a little cornflake, and that's what you're looking for."

Mary Hagedorn is a marine biologist and a research scientist at the Smithsonian Institution. She joined Penfold for the 2016 Key Largo spawning event to train scientists how to freeze coral gametes. For the last several years, she has been working to create the world's first frozen repository for corals, which includes a live frozen repository of coral sperm, embryonic cells, and small fragments.[91] "This program is important," Hagedorn wrote in a 2017 report, "because a frozen repository can keep the cells and fragments alive for hundreds of years, thus providing a hedge against extinction. These frozen resources might one day help to reseed our reefs."[92] Hagedorn has already successfully preserved frozen sperm from several coral species, including the threatened *Acropora palmata* and *Acropora cervicornis*. She also preserved, thawed, and used sperm from *Fungia scutaria* (mushroom coral) to fertilize fresh coral eggs. "Using advances in human stem cell biology," she reported, "we anticipate that these frozen coral stem cells may one day be coaxed to produce new adult coral."[93]

The cryopreservation of coral gametes presents yet another variation on the time-life-matter nexus. In this process, contemporary coral genes are frozen into the future, halting their evolutionary development to ensure that their specific genotype persists through time, albeit devoid of their ecosystems.[94] The project of "make live" here is no longer about the population, nor is it about the individual; rather, it is about the essence of life—life distilled to its bare existence—abstracted in form, frozen in time, and suspended in space.

HYBRID FUTURES

Coral, along with lichens . . . are the critters who taught biologists to understand the parochialism of their own ideas of individuals and collectives. These critters taught people like me that we are all lichens, all coral.

—Donna Haraway, *Staying with the Trouble*[95]

Whereas this chapter has focused on symbiosis, collaboration, mutualism, and fertility in all things coral, the human characters in this story have proven to be somewhat less coralated. In Chapter 3, which focused on restoration, the tensions that surfaced pitted coral scientists with more traditional tendencies against interventionists. While in many ways overlapping that divide, the frictions here are slightly different: they are based on the divergent perspectives between marine biologists and ecologists, who do most of their research in the "field," and geneticists, who mostly work in the lab. The distinction and the tensions between field and lab are not new and are certainly not unique to coral scientists.[96] Yet, as this chapter has demonstrated, this distinction is eroding, as the field (often interchangeable with the "wild") is becoming a gigantic lab, and the lab is becoming a refuge for the field.

Despite its erosion on the broader level, or precisely for this reason, many coral scientists hold on tightly to the field-lab distinction. Here, for example, is how one of the ecologists interviewed for this book responded to the geneticist's statement, cited earlier, that coral scientists romanticize fieldwork:

I noted with amusement the comments attributed to an anonymous source who criticized the lack of molecular and genetic studies at the major coral reef conference and [who] complained about the romance of fieldwork, implying that molecular and cellular studies in the lab will develop the real answers to all our problems. . . . [Such comments bring us] face-to-face with the long-running [and] never-ending campaign by reductionist biologists to do away with fields like ecology and biogeography—indeed, anything that does not begin with grinding the animal up to extract some useful bit to do lab studies with. Such people do not understand that . . . we are not going to solve the "coral reef problem" by focusing all our effort on molecular genetic studies of corals or their symbionts.[97]

Clearly, the emotional stakes are high and the biases mutual, highlighting how fractious even this small community of coral scientists is. At the same time, many of my interviewees have stressed the importance of standing united at this particularly vulnerable time. Colin Foord of Coral Morphologic, who works with coral scientists on a daily basis and was himself trained as one, reflected: "What I find to be the most frustrating thing in [the] coral science world right now is the idea that there is one best answer to saving corals, and [that] the others are wrong."[98] Foord disapproves of this approach, suggesting that "this isn't an either/or situation. We can stop burning fossil fuels *and* build artificial reefs *and* use 'super corals' in coral restoration efforts *and* ark corals in gene banks *and* monitor/preserve '50 Reefs' worldwide *and* establish MPAs/Hope Spots *and* grow corals in the classroom and underwater nurseries."[99]

As I was getting ready to submit this book for the final editing stage, I came across a new article written by eighteen coral scientists. Like Foord, these scientists point out that "sustaining coral reefs in the face of physical and chemical climate change will . . . require a multi-pronged strategy that combines intensified conventional management approaches to support ecological resilience with interventions designed to boost biological resilience." The authors end with a bold call: "Although an interventionist pathway may be at odds with the tenets of traditional conservation and its emphasis on minimalist human intervention," they write, "it has already become the necessary trajectory for both natural and cultured ecosystems." They finally warn that time is of the essence: "as species are lost to rapid climate change, so are opportunities to protect them."[100] The focus on more radical actions and interventionist pathways is no longer relegated to whispers behind closed doors; it has become a clear call sounded by prominent coral scientists and published in top scientific journals.

As befits a chapter focused on the more optimistic side of the pendulum, I will end with a few final reflections on the hopeful role of coral hybrids. Foord pointed out in one of our conversations that while the recent severe bleaching events killed off large coral populations around the world, it has also exposed the survivors—the super coral candidates. For him, then, "this disaster is a form of inadvertent 'assisted evolution' that has produced 'super corals.'" Specifically, Foord believes that Miami Beach, with its "polluted waters, man-made substrates, and discarded bicycle frames," is in fact the most cost effective coral laboratory on the planet. "Sure, spending millions of dollars on a high-tech lab that can automatically generate the water conditions of the future earth is great," he told me. "But why not start [your research] here in Miami, where those corals and conditions already exist in the 'wild'—and for a fraction of the cost?" As an aside, he added: "Weird stuff happens in Miami." It is precisely such weird stuff, according to Foord, that may have the potential to restore our devastated coral reefs.[101]

Foord concluded with an open invitation to Ruth Gates: more than anything, he would like her to join him for a tour of Miami's urban corals, as he calls them. "I want to show her the hundreds of corals cementing together the refuse of humanity, mere feet from thousands of cars that belch ocean-acidifying exhaust. After seeing them in the flesh, I think she'll agree that these are indeed the most super of the super corals!"

A Super Coral Scientist

An Interview with Ruth Gates

Ruth Gates is director of the Hawai'i Institute of Marine Biology at the University of Hawai'i. The institute occupies its own tiny island, known as Moku o Lo'e, or Coconut Island. She is also the (first female) president of the International Society for Reef Studies. Elizabeth Kolbert wrote about Gates in the *New Yorker:* "She is fifty-four, with a round face, short brown hair, and a cheerfully blunt manner ... and uses every opportunity to head into the water."[1] Gates also has a third-degree black belt in karate, which she teaches every Sunday. I first spoke with her over Skype on January 25, 2016, and later joined her for a daylong visit to Coconut Island on July 1, 2016. The text herein is an edited arrangement of the discussions we have had over the course of two years. I situated it as the last interview in the book for several reasons, one of which is Gates's central role in promoting the genomic turn in coral conservation and her vocal commitment to one of the more radical proposals in the field: assisted evolution.

IB: Why corals?

RG: I've been working on corals for more than twenty-five years (I'm getting old!) and I am as fascinated and wowed by them today, with new information coming in about fundamental aspects of their biology, as I was when I started. When I first started, I was arrogant enough to think that for my Ph.D. I would be able to solve the dilemma of how cnidarians regulate their dinoflagellate symbionts. How do they not get overgrown by the symbionts? I finished my Ph.D. in 1990 and here we are, many years later, and I still have no bloody idea! I think the corals are amazing organisms that are arguably some of the most complex on the planet. That's why I study them. All of the biological complexity [that] you find across the entire animal tree—you can find in the coral itself. I have had the luck of having been trained originally as an ecologist and then as a molecular biologist. So I work in both large and small biological scales. I like that: there are tools targeting every scale of biology [that] we can leverage to answer our questions.

My values are very different from those of many scientists I have worked with. I'm very mission-driven and that's been a tension for me with academia. That's always been the case, and when you're a female they can marginalize you, your feminine perspective, [which is portrayed as] emotional. I constantly talk about my passion for coral reefs, my emotional connection. Coral reefs are my cathedral—that's how I feel when I'm on a coral reef. I have a deep sense that this is where I am meant to be and that my skills should be applied to this place because I'm fascinated by it. Those are all wrong things for a scientist to say.

IB: How do you feel about the recent decline in reef-building corals?

RG: Well, how do you manage the need to protect with the reality that we aren't making decisions to protect? This is exactly where we are with coral conservation. To what degree do you act to protect, and to what degree do you acknowledge that nature can't do this alone? I was absolutely and completely frustrated with the documentation of declines in coral reef integrity by our community. Our scientists were continuing to say, "This is what I see, things going downhill, oh my god!"—with the inference being that someone else should be doing something about it. I took a sabbatical in 2010 and realized that there is a complete disconnect between our basic coral science and the complexity of conservation and management of an ecosystem like coral reefs. In my head, I was going around and around thinking, how do you do this? Do I even know what kind of science a conservation practitioner or manager would use or need? No, because I'm a scientist, not someone making decisions on the ground. My frustration with academia is [that] it's broken. The system is broken. Peer review is broken. We're in a highly competitive environment, competing for limited resources in an anonymous peer-review setting. Please! There is a huge gender bias in the field, and our reward system is based on silos: individual disciplines and individual contributions. Being in a silo doesn't allow you to be creative or to solve problems collaboratively, and certainly creates enormous redundancy overall.

A friend of mine, who is extremely wealthy, asked me at the time: "If you could do anything, what would you do?" It was a very provocative question and it really got me thinking. I conceived of a world where you could really work in a transdisciplinary setting: you could have a question or a problem and use that as a centerpiece for convening some of the best experts in the world with absolute relevant skills to solve it, with the understanding that time is the most important factor. It's not about individual egos at this point, it's about the mission and the questions that we would collaboratively engage in solving.

To identify the projects, I went ahead and convened that kind of workshop. I brought together a group of scientists, conservation practitioners, and managers [from] coral reef disciplines and asked them: what science should we be doing in support of conservation? Many scientists behave as if they are the brain trust—as if the people who are doing the conservation somehow aren't doing the science, they're [only] "applying" the science. I [chose] an excellent facilitator who really cut through the crap early. So right up front, we talked about the values that we bring to the table. We decided we would go on this journey of identifying projects that we all thought could be of high value. We went around the room and we could see [that] we all really needed to link and understand data much better. We have all been busy collecting data but none of us knew what our data meant or how trends related because we lacked analysis tools and the capacity to compare it with other datasets, or even an idea about whether anybody else had data relevant to our endeavor. That brought us down from the data standards to the semantics corridor: do we call the same thing the same thing in science, conservation, and management? The answer was no.

We were all interested in the way coral reefs are going to perform over time. How do we render their complexity into the conservation and management of a dynamic system like this? How do we extract information that tells us, "Well, this is something we do really well over here, but not very well over there"? Conservation practitioners are in one corner and scientists are in another—they never communicate about data. How do we do it? My lab is spearheading a large data-integration effort in collaboration with the National Science Foundation's Earthcube, which is a massive cyberinfrastructure project that integrates data across the geosciences. . . . That was one of the initial outcomes of the workshop.

At the end [of the workshop], we put the eight projects that we had identified up on the wall and we went around the room with the idea that we would prioritize them. But we soon discovered that we really couldn't prioritize: all the projects were equally important. Of those projects, many moved forward very quickly. One that hasn't gone anywhere, which was the most surprising to me, [is] water quality. It is critical that we address water quality. This gets to the massive tensions between the global stressors that are ubiquitous and the highly unique local settings. The local pressures are in some cases driving deterioration of the system way ahead of climate change stressors. Another project that came out of this [workshop] was this really ambitious project, the one we are now doing and have funds to advance, on assisted evolution.

IB: What was the most important realization you took from that workshop?

RG: Well, I realized that we haven't got time to wait for the science to tell us it's right. The time to make decisions on the ground about managing reefs is short—I hadn't been aware of that before. For instance: this week a boat grinds a coral reef down to pulp. Do I go in to restore it or not? Those are the sort of decisions that managers need to make. Perhaps the most influential thing that came out for me, before we get to the assisted evolution project, was that I started to question the 95 percent confidence that scientists all use. Managers haven't got time to wait for that shit. They made it clear that if you have an idea about something that will work and you are greater than 50 or 75 percent confident about it, they want to know. That's good enough for us, they say. Of course, it was an instantly freeing moment. I realized that herein lies the tension: it's about what we scientists think we ought to deliver to conservation managers when they're actually telling us that this level of nuance is not important. What kept coming up was that the science is failing the actual needs [of conservation managers].

As I walked away, I felt very changed. I was incredibly exhilarated by the fact that you could bring people together with a lot of thought—we set it up so that everybody felt like they had a voice. I have attended many synthesis efforts that have been really flouted by a couple of loud men who basically patronized everybody else in the room. [So] I was very careful about who I selected and got a lot of input from my colleagues.

About a year or two after that, I was back here on Coconut Island doing my thing, and Paul Allen [Microsoft cofounder and philanthropist] put out this challenge to the community: "We challenge marine scientists to give us the best ideas to mitigate ocean acidification in marine systems." So I said to my colleague, Madeleine [van Oppen], "Why don't we just put this assisted evolution project in?" I had been surprised when, during the workshop, the managers and conservation professionals thought this was something that could be acceptable, but I knew there would be pushback from the science community. We sent out a two-thousand-word essay—a concept statement that broadly explained the approach and ideas. First we were told [that] our idea had been selected to be one of the top five, and then they told us we had won the competition. Nobody could have been more surprised than Madeleine and I.

IB: Was there a prize attached?

RG: The prize was five thousand dollars each, which was nice. [But] the actual prize was the invitation to put in a full-scale proposal. That was a whole different ballgame. It's one thing to blue-sky an idea and quite another to present the nuts and bolts of what it takes to get the idea funded. It was a very large effort to put the proposal together. We were given a 3.5-million-dollar upper limit—direct, no overhead. That is an enormous amount of money to do science with. We started writing the project. It was immediately apparent to me that while 3.5 million is a lot, four [million] would do the project better. I went to the foundation with this request and they came back and said fine. Perhaps having a British accent and being able to sell things helped.

It was quiet for a long time. Then the foundation came back to me with all these questions about what we wanted to do. It took many rounds of questions. . . . At the end of the day, it went really well and they presented it to Paul Allen. Then they said they're definitely going to award it, but are going to award at 2.7 million and we would need to get a 1.3-million match. This was eighteen months into the process. Everybody was like, "you must feel great, you got 2.7 direct, that's a shitload of money." I replied that I do feel great but that we can't do the project, as written, on 2.7 and, frankly, finding a 1.3-million match is like climbing the Everest—it would take me twenty grants to get that. Had we been told earlier about this matching requirement, I would've been okay with it, but it just didn't feel doable at that late stage. So, without the permission of my administration, I went back and said, "No. I can't do it." They eventually came back and said we agree with you, we're giving you the full four million. In part, I think it was because we had been so positive through the process. [So] when I really put the boundary down, they believed it. Frankly, four million is transformative, by any measure, for this kind of work.

IB: What kind of work are we talking about—can you tell me a bit about the nuts and bolts of this project?

RG: It's a really ambitious endeavor. There are three clear things that we're doing in Hawai'i, and an additional thing that we are doing in Australia that leverages the Australian corals. The project is a proof of concept that we can raise the performance thresholds of corals to stay abreast with the rates of change in climate change stressors. The idea is that we can potentially breed corals [so that they are] prepared for the future, [instead of] failing in the face of this future. This gets to another piece: how do you nuance something like this so that it is scalable across reefs, which go from thirty species in Hawai'i to over three hundred on the Great Barrier Reef, or, in the opposite direction, to the Caribbean? What are we doing? I would argue: we're identifying specific things that work for particular types of corals.

The word *types* is the piece people don't fully understand. We're interested in functional groups: big boulder corals—incredibly important to diminishing storm energy—versus branchy corals, [whose] primary role is to create space for other organisms to live and are pivotal to food security. You have structural integrity versus biodiversity maintenance in those functional groups. Those groups are also biologically different: whereas the big boulders are slow-growing, have deep tissues, [and] tend to have one symbiont type [see, e.g., figure 41], these really entrepreneurial branching corals with these gorgeous forms are fast-growing, tend to have more than one symbiont in their tissues, often have tissues stretched over the surface, [and] are probably not as long lived.

IB: What is the overall idea of the project?

RG: Really, the basis of our project is that we're assisting nature, we're accelerating natural processes. We're not creating anything new: we're doing what nature does, and just trying to find ways to do it more quickly. If we take the

FIGURE 41. *Porites compressa* is an important reef-building coral in Hawai'i. Gates Coral Lab scientists fragmented twenty genotypes from this species in their lab on Coconut Island for a reciprocal transplant experiment (see figure 42) that uses natural variation to investigate coral physiology and symbiotic relationships. Photo by Shayle Matsuda, September 2016. Courtesy of Gates Coral Lab.

[corals] performing well now, can we lift their biology to perform even better? Can the best performers now become the better performers of the future? We're doing that, for example, through selective breeding—just classic selective breeding. We're interacting very hardy genotypes. And then we're going through conditioning—running our corals on environmental treadmills— with the idea that stress exposures that don't kill you make you stronger. . . .

People ask: what does it matter if you can bring a few corals into an environmental treadmill in a lab, how can that help the reef? And I [answer] them: what if we understood well enough that if you raise the temperature two degrees for a week, the majority of corals survive, and the next time you expose them to this same level of stress, they simply don't see the exposure as stressful? Why couldn't we go out with an engineering solution and do that at the scale of an entire reef? In my mind, that's absolutely plausible.

We feed at least six billion of our seven billion people. That's an astonishing scaling up. Nobody would've imagined you could produce enough food from the land to do that.

Here is a different scenario, and this exemplifies how differently coral reefs are doing in different places. There is a place in Hawai'i, Kaneohe Bay. People say it's not like any other place in the world. Guess what? Raw sewage was poured into this bay until 1978. Almost every single coral on the bay was dead. The bay was covered in fleshy algae—it was a green bay. They diverted the sewage and, since then, the reefs have recovered and we have almost 100 percent cover on some reefs now. An extremely resilient reef system has emerged, built from corals that survived the intense stress of the past.

There is also a reef in Taiwan that is exposed to sharp variations of six to nine degrees Celsius a day because of hot water effluent from a nuclear power plant and to an upwelling of cold water at night. It's flourishing! What I think happened is [that] when the power plant was turned on, that killed almost all the corals there. The ones that survived are extreme performers in that community. Now, that whole reef is proliferating and even extending beyond that site. What you've got there is a fountain of extraordinarily hardy individuals. They've released their eggs and sperm. They are free to be fertilized by any eggs and sperm in the environment, from reefs around the corner—say, from corals that have not been exposed to those [adverse] conditions.

IB: This is very interesting. What are the central criticisms directed toward your project?

RG: People will often say, "Oh, you're going to narrow the diversity of coral genetics." That may be true. [But] let's face it, climate change is the most radical narrower of genetic diversity across all living systems in the planet. We have to step back and ask, "What about the relative risk of these approaches?" If we do nothing, where is the endpoint going to be? I can pretty much guarantee that even a 1.5°C increase is going to kill a lot of corals. I am realistic about where we are and about what the trajectory of change for coral reefs is. For the most part, they are almost all declining. That means it's not sufficient to step back and wait. People tell me that nature might adapt by itself. It might, but it doesn't seem to be; not quickly enough at least. And what if we do step back and nature doesn't [adapt]?

The complete annihilation of reefs by climate change would be an appalling outcome.

Our project is about proof of concept—showing, for example, that yes, you can selectively breed corals, although they are not easy to breed. We've learned a lot of things already. Corals don't like their gametes to be touched at all. People were originally taking eggs and sperm and trying to get them to reproduce in tubes; corals hate it. They need to interact passively. So we designed these big arrays with nets to separate corals and control the way sperm moves directionally into different compartments. But it's not easy to do that. That's all done here on Coconut Island. In Australia, we are working at the National Sea Simulator, which is a really a massive aquarium facility. There are replicate tanks, and you can keep the environment attuned to whatever you want it to be. Here in Hawai'i, we have exactly the same facility except it's way more rustic. We are pumping water directly out of the bay into our facility. We're exposing our corals to everything that is in the bay.

IB: So what are you working on at the moment?

RG: Right now, we're doing a massive experiment with twenty-five genotypes that are growing in multiple types of environmental conditions. We're working with both acidification and temperature. Since both are changing together, we don't see the logic of separating them. What we do, basically, is heavily replicate our conditions and ambient controls. For example, we have a bank of twenty-four large tanks that can each hold multiple fragments of those genotypes we are testing. The goal is to collect eggs and sperm from those genotypes and selectively breed those so they get acclimatized right now [for] the conditions of the future: warmer and more acidic. To be able to do it right, statistically, you have to design independent treatments. We are doing it right.

Madeleine is doing exactly these same experiments on different species in Australia: gametes with the same genotypes are developed in tanks with future conditions or with ambient conditions. When we actually do selective breeding, we can distinguish with the propagule whether or not we see selective advantage as a result of these conditions by reexposing them to warmer and more acidic waters and assessing their responses. Do they do better or worse? Do the offspring of conditioned corals do better than the ambient and does the offspring do better than the parent? Have we actually facilitated a positive acclimatory response through our experiments? Here,

FIGURE 42. From left to right: Gates Coral Lab technician Josh Hancock, Ph.D. student Ariana Huffmyer, Dr. Katie Barott, and Ph.D. student Beth Lenz work underwater on a reciprocal transplant experiment in Kaneohe Bay, Hawai'i. During the experiment, the team moved 3,200 coral fragments of two of Kaneohe Bay's most abundant reef-building coral species, *Montipora capitata* and *Porites compressa*, between two sites with varying pH regimes to observe how this transplant affects coral physiology, symbiosis, and reproduction. Each rack on the large metal structures carries coral nubbins from the same parent colony with the same genotype, and each nubbin is attached to an acrylic plug with a unique identifier engraved on the side. These researchers are randomizing the genotype locations on each of the experimental racks. Half of the corals will stay at this site, and the other half will be moved to another site. Photo by Shayle Matsuda, September 2016. Courtesy of Gates Coral Lab.

we are essentially interweaving two things: the conditioning, or exercising, and the selective breeding.

At the end of that experiment, we will also try what I believe will be potentially the most difficult thing to do: a slight tweak to the symbiosis [by] introducing what we know is a symbiont linked to high tolerance. We are working with a symbiont that is very closely related to the native [one] and in the same big group, but we always find it in corals that rely very heavily on the symbiont for food, which is why we refer to it as the best feeder. Its official name is C15 and it is also found in the most environmentally resistant corals in the world, like the massive *Porites* species. C15 is an extraordinarily hardy symbiont. We're trying to introduce it into a coral called *Montipora capitata*, which also has a good symbiont called C31. We are doing very close manipulations with the idea that, immunologically, it makes more sense to try and introduce something as similar as possible to the one you acquired first. The question is, will the coral take it up? I believe that the early life-history stages of all organisms are more permissive. What we're attempting to do is to introduce [this symbiont] in baby corals and see if we can close the door to other symbionts so that when the coral has it, it has it. And then we will test whether the corals grow more quickly and tolerate more stress in the environment. We don't know the answer; it's never been attempted.

So, in essence, there are three parts to our project. One is the conditioning, [which is] coupled with, and followed by, selective breeding. Finally, with the offspring we are attempting to manipulate symbiosis. We are trying to do all three things on the same corals. Our argument is [that] if you can do three things instead of one, maybe you can raise stress tolerance further. It's quite an elegant design but a monster to reverse-engineer. That's exactly what we did: we said let's go right to the end and then let's think we're a year or two from there—what would we have liked to have tested during this period? We reverse-engineered our experiments to figure out the number of tanks we need, the number of replicates we need, and the sorts of lessons that we can learn along the way. We have a whole heap of corals that will undergo the same set of experiments but [that] are not manipulated at all—[these are] the controls; they will be our brood stock and will be used in later experiments as well.

We plan to get through two generations with one of our brooding species. We are also doing an experiment this year with a spawning species. We will be repeating the conditioning experiments for that species, too.

IB: And what if something kills your experimental corals?

RG: We have extras; we have to keep extras just in case we have a disaster. We over-collect so that if we have any deaths in our tanks, we can basically [replace those corals]. We've got them all in a common garden, essentially: all in the same big tank, all exposed to the same recovery. We're just waiting and monitoring their recovery.

We've done previous work demonstrating that if you take a coral and hit it with a hammer, for about two weeks it doesn't perform normally at all, its physiology is all off. And whose wouldn't be? It's just been hit with a hammer! [So] we have metrics of normality that we want the corals to reach before we begin the next step along the experimental pathway. Through our experiments, we're monitoring everything we can so that we get a sense of what each of our genotypes looks like. Some are more active than others and grow faster than others. By tracking genotypes, we can now start to understand the variations in the system. What does it mean if you grow more quickly—does that mean you actually put out less reproductive effort? Do we want to get them really well fed before we move to conditioning? [And] what do we have to do with our corals to get them to be in the best state to be conditioned? At the end of the day, is six months at a one-degree Celsius change rate the same as two weeks in a two-degree rate? Can we induce the same kind of effect?

IB: What about genetics?

RG: Our goal is to get to a biological capacity that could be used downstream for climate change adaptation and coral reef restoration. As part of the project, we are sequencing the genomes of five Hawaiian corals. Our goal is not only to do the experiments and know that we can get there, but also to figure out how we got there. I want to understand the basic biology that underpins the processes we are attempting to accelerate. For the conditioning experiments, we're looking for the presence or absence of indications in the genetic material that epigenetic change has occurred, and whether or not this change is retained in the offspring. There are signals within the genetic information that tell you that.

We will never relay the details of the DNA work to a manager—there is simply no need to go to this level of detail. But I'm an academic and I'm excited about furthering our basic understanding of ancient metazoans. It's about balancing applied and basic research and knowing that the applied science will be ten times more valuable with really rigorous science validating the approach behind it.

But at the end of the day, this is not just an academic project. It's a project driven by an unavoidable need: we [must] get there before the biology goes [down the drain]. It's easier to criticize than it is to embrace the idea that maybe we are here: we are hitting this critical point in time. There's no doubt [that] we're already doing that kind of work in forest restoration. So [we need to] talk about things they've done well and things they've not been able to do very well. What are the best lessons we can take from the forest restoration people—from the cottonwoods, for example? How do we accelerate our field as quickly as possible? Rather than just looking within our field, we're going to the very best practitioners in other fields.

IB: What is the role of genetic modification in your project?

RG: The first article on our website was the "designer reef" article, which was published as a comment in *Nature*.[2] The word *designer* made people say, "OMG, GMO for corals!" That's what came back to us. But that's not what we are doing, we're not doing GMO for corals. We're doing what nature has done all the time, but just can't do it quickly enough. People are really uncomfortable with GMOs, although, let's face it, it's already being done in so many places. We've already crossed over into engineered biology in the food supply. And we have engineered our land. Yes, engineering is coming and we should talk about the relative risk of engineering versus doing nothing. That's always my argument. But we should start with the least risky approach first—and that is what we are doing.

The timeline for me is pressing. I don't know what the outcome of our project will be. I have no idea whether it will be successful. But [in a deeper way,] we're already successful. Because we're sending out a more hopeful message, something we can do actively. I'm using this project as a platform for a variety of things. One, let's get some capacity in place to think about how we will do it if we need to do it. Two, let's do that before the biology is dead. It's much bigger than one discipline, so it allows me to address my own intrinsic disgust for siloed science and to promote collaborative endeavors. It also allows for a conversation about why we as scientists have not done a very good job at articulating to the majority of people that these are critical ecosystems. [How do we] get people to be enthusiastic about what corals are?

The positive message is definitely helping. People want to do something, they want to help. The gloom and doom is paralyzing. The scope of the climate change issue is paralyzing. I keep saying that climate mitigation is the critical other end of the stick, but we have to be thinking about adaptation if mitigation doesn't work. And even if mitigation works, there's going to be a lag. We've got to be thinking more strategically now, so [that] we are ready.

I understand that some scientists have spent their entire careers hitting people on the head with climate change and with the need to mitigate fossil fuel burning. I get it. But criticizing us because we are doing something different seems unwise. It's a joint affair. Are we going to be able to restore hundreds of thousands of kilometers of reef? I don't know. Are reefs going to be this boutique oasis of what life used to be? Maybe. We've got to be realistic about where we are.

We may not know what the solution is or even be able to predict the outcome, but we can at least be in a conversation and come to that conversation with an enormous scientific knowledge about what corals are. I'm sickened by the statement "All corals are going to be dead by 2050." It's a lie, a dirty lie. They're not going to be dead by 2050. Massively degraded, perhaps. But that's a very different statement. The discussion among coral scientists has been so academic, and the messages that make it out to the public are so alarming, with no sense of hope. Let's talk about the real risks of intervening and let's look at the risk of doing nothing. The second risk is so much bigger. The practitioners see it very clearly. But they don't have the capacity to advance a discussion on intervention or to develop corals that can survive the future—scientists do. We've got to stop telling everyone that reefs will all be dead by 2050! People are walking away.

Conclusion

Coral Scientists on the Brink

What do you do when your world starts to fall apart? I go for a walk.
—Anna L. Tsing, *The Mushroom at the End of the World*[1]

CORAL INSTALLATIONS

In 2005, two Australian sisters, Margaret and Christine Wertheim, began to crochet a woolen reef in their living room. They dedicated this crocheted reef to the Great Barrier Reef ("as a homage to the Great One"). A few years later, this unique fusion of art, science, mathematics, and activism became the largest community art project in the world.[2] So far, more than eight thousand people in twenty-seven countries, mostly women, have collaborated to crochet displays of coral reefs.[3]

The inspiration for making crochet reef forms began with the technique of hyperbolic crochet, a form of weaving that utilizes a surface that has a constant negative curvature (think lettuce leaf), constructed in 1997 by a mathematician from Cornell University. The process relies on a simple algorithm pattern, with variations (mutations) on this algorithm introducing permutations of shape and form. The Wertheim sisters adopted these techniques and elaborated on them to develop an entire taxonomy of reef-like forms. Coral forests, bleached reefs, and biodiverse reefs have been woven into existence with a variety of materials, including wool, cotton, plastic bags, and Saran wrap.

FIGURE 43. Mushroom disk corals (*Cycloseris* spp.) are large and solitary corals from the Indo-Pacific. Unlike many other corals, they live unattached to the reef in adulthood. When buried in sediment, coral species in this genus can use hydraulic pressure to flip over or reemerge. Courtesy of Coral Morphologic.

Through this ongoing evolutionary experiment, a worldwide community of "reefers" has been bringing into being a growing crochet of life.[4]

Making art on a polluted planet feels like a "looping of love and rage," writes Donna Haraway, a feminist scholar of science and technology studies, who has been involved in this project since its early days. Reflecting on the project's connections to the themes of hope and despair, she writes: "Neither despair nor hope is tuned to the senses. . . . Neither hope nor despair knows how to teach us to 'play string figures with companion species.'"[5] I agree with Haraway's assertion that the hope-despair extremes, wedded in their ever oscillating dynamic, are hardly a sustainable model for our relationship with corals, and with one another, in the Anthropocene. I have therefore argued throughout this book for alternative—more active and sustainable, yet still dynamic and relational—forms of hope. More than merely despair's opposite, such alternative forms of hope in fact emerge from despair and contain it. Although most of the coral scientists I spoke to didn't call them by name, they knew and experienced such hopes, telling me about the different paths they have pursued that empowered them to continue their work despite the daunting tasks and the low odds of success.

Through crocheting, corals have thus woven together, and inspired, those who care for them. Corals have also provided inspiration for digitalized science-art forms.[6] Coral Morphologic, mentioned in previous chapters, is a Miami-based underwater multimedia endeavor founded by Colin Foord and Jared McKay in 2007. Premiered at Art Basel Miami, their 2010 night installation *Artificial Reef* showcased a local coral reef projected on prominent skyscrapers in Miami Beach. The limestone that constitutes these buildings is mainly composed of the pulverized fossils of coral that colonized south Florida when it was submerged in the ocean some 125,000 years ago, Foord and McKay explained in a video.[7] They emphasized that

> projecting the corals onto those buildings references the geologic past, it references the technological present we're in, and [it references] the potential future, where if sea levels continue to rise, corals would have no problem coming right back in and cementing themselves to themselves. . . . The whole city is a giant bed of coral that used to be underwater and might go back under.[8]

Enchanted by the shape, color, and fluorescence of local coral species that survive and even thrive in the polluted waters of Miami Beach, the artists have been cultivating these corals in their own basement as a form of living art. By placing images from Coral Morphologic's installations at the head of each chapter, this book has celebrated corals as revelatory figures with whom we may think through, and feel, our contemporary social and ecological vulnerabilities. "Recognition of dying coral reef ecosystems in warming and acidifying seas was at the heart of advancing the very term Anthropocene in 2000," Haraway points out.[9] The corals' otherworldly beauty still serves as a fleshy reminder of what is at stake for nature conservation in the Anthropocene.

Not unlike artists, modern coral scientists have been inspired by the corals' striking aesthetics and have come to realize just how interwoven all forms of life are.[10] This, I believe, is what Lorenzo Bramanti meant when he spoke about the mythological *Corallium rubrum*,[11] what Katia Chika-Suye referred to in her spontaneous appreciation of the endemic blue rice coral ("Isn't it so beautiful?"),[12] and what Ruth Gates implied when she referred to corals as her "cathedral."[13]

Conservation, in fact, is a form of art—here, the art of cultivating both corals and hope for the future. The corals themselves provide the strongest sources of inspiration for such hope. As one of the earliest instances of symbiosis recognized by biologists,[14] corals are a constant reminder of the importance and pervasiveness of collaboration and mutualism. They have thus been central to the recent scientific realization that "we have never been individuals" and that "we are all holobionts"[15]—assemblages of microbial forms of life with complex interrelations. "But what happens when a partner involved critically in the life of another disappears from the earth? What happens when holobionts break apart?" These kinds of questions have to be asked, Donna Haraway tells us, "if we are to nurture the art of living on a damaged planet."[16]

Many of my interviewees are constantly awed by the levels of complexity exhibited by these supposedly simple organisms that we generically call corals. *Coral Whisperers* has illustrated how this complexity and its ecological significance have recruited scientists from around the planet to become spokespersons for these creatures. At the same time, it has documented the frustration experienced by many of these scientists as their ardent speech was relegated to a whisper: either unheard or outright ignored by policy makers and the greater public.

BIPOLARITY

Alongside the story of coralation, complexity, and wonder, this book has also told another story—about the polarized state of the scientific community that studies corals. Struggling to deal with the crisis of massive bleaching and with the accelerated death rates of reef-building corals, coral scientists have been oscillating between hope and despair. The book's structure has mirrored this oscillation: from documenting the depth of depression in response to the third and most recent global bleaching event, to witnessing the hope in restoration projects; from the frustrations with law's failure to regulate the global threats to coral populations, to future-oriented experiments in assisted evolution. The movement of this pendulum has provided not only a structure but also a timeframe for this book. Tick-tock, the pendulum persists, a goad for the increasingly urgent work of coral scientists. The fast pace of coral decline is mirrored in this chronology by the work of the coral scientists, who are ceaselessly trying to catch up.

I have argued that the hope-despair polarity maps onto the coral scientists' internal debates over the naturalness and pristineness of coral reefs and about the appropriate levels of intervention in their ecosystems. The community of coral scientists is split on this issue. On one side are those who believe that the intense human impacts on nature call for similarly intense conservation measures. According to this position, more interventionist technologies, such as massive restoration projects and assisted evolution, are justified. If there was ever a time for radical and even experimental actions, these scientists contend, it is now. On the other side are those who argue that human hubris and interventionist approaches are what have caused the current impoverished state of our planet in the first place. In their view, human-induced climate change has already irrevocably damaged coral reefs, and no local or technical fixes will reverse or even halt their mortality. The only truly effective thing we can do, according to these scientists, is to cut greenhouse gas emissions on a global scale, and we have to do this now. The debate over the efficacy of coral restoration and its place within coral conservation is very much entangled in such underlying viewpoints, all searching for the right balance in the management of a radically transforming nature.[17]

Finally, as I have realized during the course of writing this book, coral scientists struggle with a double-edged sword. To put it simply: if humans can fix nature, there is no reason to change our damaging ways; at the same time, if all is hopeless, there is still no reason to change our ways. Preservationists criticize interventionists for thinking humans will develop new technologies to save us; interventionists criticize preservationists for being too pessimistic about human nature. The coral scientists I have spoken with are trying to figure out both how to work from a place of hope without losing sight of their limited capabilities and, simultaneously, how not to fall into despair without exaggerating such capabilities. In the words of one climate scientist: "There is no need to exaggerate the problem of climate change, it is bad enough as it is."[18]

CORAL CLIMATES

Coral Whisperers is a book about bearing witness. Specifically, it is a testimony of how coral scientists are coping with the unfolding coral crisis of our time: the most catastrophic period for reef-building corals in human history, and for the preceding sixty million years. Corals are not only canaries in our global coal mine, visible manifestations of the radical changes we are causing in our ecosystems and cultures; they are also a bellwether, sounding the way through this mess toward a more mutualistic way of being in the world that resists classification, linearity, and binary thinking. And while their acute bleaching in multiple sites across the planet signals that something has gone terribly wrong on a global scale, the multiple and varied ways that scientists, managers, and the corals themselves have been dealing with bleaching, and the science-law coralations that have evolved in the project of caring for corals, remind us that the situated, the local, and the relational may still matter.

Looking at how the scientists have been interpreting the deteriorating state of corals and at how they have narrated it to the world, one recognizes that, above all, these scientists are human. As trite as this may sound, scientists, too, have personal intrigues and professional debates, as well as distinct personalities that often determine their scientific tendencies. One also realizes that (with one exception, which I will return to shortly) coral scientists are anything but unified, especially when it comes to dealing with climate change.

While some scientists argue that coral conservation must make climate change its number one priority, others feel that the elephant in the room has already trampled over everything else and caution about its dominating nature. "Everybody blames it all on climate change," coral grower Ken Nedimyer told me a few days after Hurricane Irma hit his nurseries.

> It's kind of a cop-out. The problem with blaming everything on climate change is that people throw their hands in the air and say [that since] they can't stop climate change they won't do anything. But if we address things that we can address, such as overfishing, point-source pollution, and runoffs, then we can change things—we can make it better.[19]

Similarly, marine biologist Jeremy Jackson observed that for many local conservation managers, climate change has become too easy of an excuse for doing nothing. In his words,

> If it's all those goddamn gringos in the north that [have] made things bad, then I don't have to do my job, I don't have to regulate fishing, I don't have to tell the hotel people that they can't build a hotel right smack on the shore, or that they don't need a golf course with all that fertilizer in [one] tiny little spot. The coolest thing that we found is that the places that were protected from overfishing and runaway coastal development were a lot more resilient to climate change.[20]

According to geographer Mike Hulme, the heightened attention to climate change has resulted in "climate reductionism."[21] Historian Dipesh Chakrabarty, who is also concerned about its reductionist properties, insists that climate change should be understood as part of a broader complex. Rather than merely an issue of emissions or a result of the inequalities of capitalism, climate change should be viewed "as part of a complex family of interconnected problems, all adding up to the larger issue of a growing human footprint on the planet."[22] How to address human-caused global warming and ocean acidification, and what prominence to give these occurrences when conducting coral science, is one of the central disputes facing coral scientists on the brink.

After all is said and done, though, and despite the tensions and disputes, the community of coral scientists are united around one thing: their love of corals.

SOME SAY LOVE

Like scientists in many other disciplines, many coral scientists don't want to seem emotional about their work. "Why do you love corals?" I asked dozens of my interviewees. "What is your favorite coral species or site?" Initially, I was frustrated with the answers I received, mostly rational descriptions of corals as important ecosystems and service providers. I can't count the number of times I've heard or read that "reef coral ecosystems sustain services valued at around $10 trillion per year, are home to over a million species, and feed and support the livelihoods of hundreds of millions of people."[23] While one can't overstate the importance of these functions and services, I wanted to hear something more intimate, more revealing of the bond that the scientists form with the particular corals they study. When I was asked to talk about corals to my university's alumni in 2016, I invited one of my colleagues, a prominent coral scientist, to join me. I would ask him questions, just like we had done in our interview, only this time in front of a larger audience. It is important for lay people to hear about corals directly from scientists like you, who care about them, I finally convinced him. Setting the stage for more personal reflections, I kicked off the interview with "Why did you choose to dedicate your life to corals? What is so special about them?" But he wasn't going there. His response: "Well, nothing is uniquely important about corals; my fascination is more with their function as an ecosystem."

At the time, this answer caught me off guard. Having interviewed dozens of conservation biologists who work with terrestrial species such as rhinos, gorillas, and even toads, and who felt pretty comfortable telling me about their affection toward these animals, I found this response, and many similar ones from other coral scientists,[24] to be oddly unemotional. And yet, as I have documented throughout this book, these scientists do experience an intimate relationship with corals. And they are now finally coming to realize that intimately whispering to and listening to their beloved corals is not enough; if they want anyone to listen, they must convey the message of corals to the rest of the world, and they must do it in a passionate, rather than a detached, way. For many coral scientists, it was easier to speak about their emotional relationship to corals through visual aids, such as images. I have made a point to include in this book such "unscientific" reflections alongside the images that triggered them (see, e.g., figure 44).

As this book has shown, questions of intimacy are also tied to more conceptual questions, such as what is an individual coral and which lives matter. In her office at Boston University, doctoral candidate Liz Burmester lowered a coral skeleton from a shelf to show me. "This is the northern star coral, the northernmost stony coral species in the United States," she said proudly, as if introducing me to a family member.[25] This temperate-zone coral lives just outside of Boston, so Burmester doesn't have to travel to faraway tropical islands to retrieve it (something that many other scientists do readily, especially during the cold northeastern winters). What intrigues her most about this coral species is its ability to display both symbiotic and nonsymbiotic traits—even within the same colony. Such species are called *aposymbiotic*, she instructed me, and I diligently added yet another word to my coralated dictionary.

During my visit, Burmester was running a battery of genomic tests to figure out how aposymbiotic corals manage this split identity, and was

FIGURE 44. Ángela Martinez Quintana, a Ph.D. student at the University at Buffalo, told me: "I collected these soft corals (*Paramuricea clavata*) in Cap de Creus, close to Girona, Spain, in May 2010, to perform an experiment as part of my master's [degree] in marine science. They are endemic to the Mediterranean Sea, and are major ecosystem engineers on the Mediterranean's rocky shores from fifteen to four hundred meters' depth. I am kissing these corals in the picture because I created such a strong bond with them. I learned so much from them, and I am where I am now thanks to them. Working with these corals was the best year of my life—a dream that came true. I built the aquaria system from scratch. Everything needed to be perfect. The day I collected them, I needed to limit the amount of stress for the corals, so I took them from the sea, at twenty meters' depth, and transported them in a cooler to the Marine Science Institute of Barcelona, where I was doing my master's, all in less than eight hours. I maintained the corals for several months—males and females. They were very delicate so I put in a lot of effort trying to keep them happy, and waited for them to spawn. I controlled the seawater parameters constantly, and made sure that they had enough flow and that the water got renewed every four hours. I fed them twice a day, cleaned them, and controlled the water temperature. I even had to run to the lab a couple of times in the middle of the night because the water pumps stopped working or the temperature got to be too high. I loved it! The corals spawned in June 2010. I was the happiest person in the world when that happened. After the experiment, I transplanted them back to the ocean. I did not have the time to publish anything about that experiment yet—I started another experiment with larvae, and then another one and another. I am still trying to find the time to publish that first experiment with my beloved coral. Hopefully, 2018 will be the year" (e-mail communication, June 17, 2017). Courtesy of Ángela Martinez Quintana.

grinding up their tissues for these tests. In a conversation we held later, I asked several members of her lab how they felt about the routine procedure of grinding up their animals. "We're not really killing the animal, because the genotype is still there," they told me, explaining that their actions best compare with cutting a branch off a tree. Such actions are especially justified when performed for the benefit of the entire species, they all agreed. I wondered if they would react similarly if instead of coral polyps we were to talk about puppy triplets or even human twins. Colin Foord of Coral Morphologic raised a similar question when I told him about Gates's coral acclimatization experiments. "This form of assisted evolution is often effectively a 'torture chamber' for corals," he stated emphatically, adding, "It is important to remember that corals are animals, too!"[26] These, precisely, are the types of ethical questions that arise when one works with, and cares about, corals.

LIVING ON CORAL TIME

Alongside the intimacies of working with corals, *Coral Whisperers* has also explored the importance of *thinking* with corals. Corals challenge our dualistic thinking, upsetting many traditional binaries that are ingrained in modern perceptions of the world. In addition to challenging the distinctions between life and death and the very concept of aging, corals undermine basic human assumptions about the relationships between plant and animal, individual and colony, fixed and mobile, male and female, hybrid and pure, sexual and asexual reproduction, and organic and inorganic matter. More broadly, corals challenge our all-too-human definitions of nature and our natural assumptions of what it means to be human. Finally, they challenge us to rethink how we might more wisely govern life, be it terrestrial or oceanic, and how we might better care for our coralated communities and futures.

Thinking and being with corals also raises important questions about time. During my work on this book, I encountered a pervasive anxiety with time and its measure on the part of my interviewees, what art historian Pamela M. Lee refers to as "chronophobia."[27] Relatedly, philosopher of history Reinhart Koselleck characterizes late modernity as being a "peculiar form of acceleration."[28] These theoretical references relay something of the temporal crisis that is a signal feature of the Anthropocene. "And like all good crises," Lee writes, this one, too, "calls for a certain degree of decision making. How is one to act in the face of it?"[29]

The notion of time is central to the story I have told here. Throughout the book, a feeling of looming catastrophe contributes to the sense that there is no other time to act but the present moment. The realization that time is of the essence has transformed many of the coral scientists I have spoken to into political advocates, even as many of them feel that their urgent call for action has been falling on deaf ears.

How much time do reef-building corals have until they bleach once again on a global scale? How long until their ecosystems collapse, or until they become extinct? And how entangled will this extinction be with the death of oceans—and of life on earth—as we know them? Predicting the exact year by which reef corals will die out is akin, in many ways, to pronouncements about how many minutes are left until doomsday. As it happens, the atomic Doomsday Clock, which has come to signify this apocalyptic mode of thinking, was expanded in 2017 from its sole focus on nuclear annihilation to include climate catastrophe. As the Science and Security Board of the *Bulletin of the Atomic Scientists* explained,

> In 1947 there was one technology with the potential to destroy the planet, and that was nuclear power. Today, rising temperatures, resulting from the industrial-scale burning of fossil fuels, will change life on Earth as we know it, potentially destroying or displacing it from significant portions of the world, unless action is taken today, and in the immediate future.[30]

Climate change has been assigned the status of a nuclear war, pushing the dials of the doomsday clock that much closer to midnight. Increasingly, this temporal catastrophe is coming to govern all forms of action or inaction by coral scientists.

In the face of this perceived race against time, some coral scientists are working to accelerate coral evolution. In their quest, they have been looking back in time to learn about prior climatic shifts and to study how corals have evolved under those conditions. The rings secreted onto the skeleton of corals as signatures of annual growth, as well as the DNA of their living bodies, are fleshy archives through which scientists "read" grand temporal trajectories stretching millions of years into the past. Time also extends from present to future, this time through applying climatic models to anticipate warming and its impacts on corals in fifty-year intervals up to two hundred years ahead.

The role of time and its interplay with action has assumed many shapes in my interviews. "We are all interested in the way coral reefs are going to perform over time," Ruth Gates told me. The central question, according to Gates, is how to render the complexity into a management plan that would tell us how to act. "We haven't got time to wait for the science to tell us it's right," she warned.[31] Israeli marine biologist Baruch Rinkevich cautioned, similarly, that "buying time will not suffice for the reefs of today to transition into the reefs of tomorrow."[32]

The temporal crisis, monitored and predicted to the finest detail in the coral context, has become—for Gates, Rinkevich, and many other coral scientists—an urgent call to action. "Really," Gates told me, "the basis of our project is that we're accelerating natural processes. [We're] trying to find ways to do it more quickly." Present and future are interwoven in experiments that examine whether the best performers today may become the better performers of the future. "It's easier to criticize than it is to embrace the idea that maybe we are already there," she said. "We are hitting this critical point in time. Time is pressing." For Gates, then, urgency is intimately tied with action. "Pointing to something we can do actively provides a more hopeful message," Gates believes. The whispers of corals, and the echoing whispers of coral scientists, are at last crystallizing into an urgent call for action.

CHASING CORAL: AN AFTERMATH

As I was finishing this book, the award-winning documentary film *Chasing Coral* was released.[33] Produced by the same team as the documentary *Chasing Ice* that was released five years earlier, *Chasing Coral* engaged many coral scientists in its production and was much anticipated by many in this community. Settling around the screen with my two daughters and several colleagues from the Rachel Carson Center in Munich, where I had just spent a very long spring and summer writing this book, I was excited and hopeful: the stars seemed to be aligned for this movie to provide me with just the right conclusion.

We all watched intently as the documentary unfolded. Although many coral scientists—most of whom I already knew from my research for this book—flashed on and off the screen, the movie's real protagonists were nonscientists: Richard Vevers, the former-toilet-paper-salesman-turned-director of the Ocean Agency,[34] and his fabulously nerdy technician, Zack, also a coral hobbyist in his spare time. Although initially baffled by the filmmakers' choice of characters, later I came to understand it much better. While it concerns itself with the fate of corals, *Chasing Coral* is fundamentally a movie about climate change. And because coral scientists have been unsuccessful, thus far, in effectively communicating the connection between coral disaster and global warming to the larger public, it was now up to others to do it.

Chasing Coral is a sad film. Together with the protagonists, the viewer is bombarded with disturbing imagery of coral bleaching from around the world, and especially from the Great Barrier Reef, where corals hundreds of years old disintegrated into algae-covered skeletons over a period of four months. This process is deftly captured by the camera lens, and a number of coral scientists interpret these devastating scenes. "If coral reefs are lost, we're affecting the life of a quarter of the ocean," one remarks. "It's easy to think about the fate of an individual species," says another. "But what is a little harder to explain is the beginning of an ecological collapse of the entire ecosystem. . . . We're talking about the possibility that entire classes of organisms would go extinct." Ove Hoegh-Guldberg clarifies this: "When scientists say they're researching climate change and coral reefs, it's not about whether climate change is happening or not. It's really about the uncertainty of if it's going [to be] bad or *really* bad." Ruth Gates then summarizes: "Coral reefs will not be able to keep up, they will not be able to adapt. And we will see the eradication of an entire ecosystem, in our life span. That is a very gloomy statement. But unfortunately it is true." From this point on, the scientists' statements and questions are hurled toward the viewer at an accelerating pace: "Everything in our planet is connected. What we are doing is pulling out the card named 'coral reef,'" one voice states; another asks rhetorically, "If we can't save this ecosystem, are we going to have the courage to save the next ecosystem down the line?"; and a third inquires, "Do we need forests? Do we need trees? Do we need reefs? Or can we just sort of live in the ashes of all of that?"

Hope finally arrives in the person of "Charlie" Veron, an elder of the coral science community who humorously refers to himself in the film as a "gloomy old man." Zack idolizes Veron. Toward the end of the film, he visits Veron's home to seek his perspective about the changes he had witnessed over the past forty-five years. Veron reflects: "Then, it was a totally different mindset because the reef was [going to be] there forever—there was no question about it. You even wondered why it would need protection, it was so big." But now that's changed, he states wryly. "I'm glad I'm not your age," he tells the twenty-something Zack, explaining that he could not bear to witness the most beautiful and beloved thing in his life, the Great Barrier Reef, getting trashed. Yet he concludes in a different tone: "I've been diving for forty-five years of my life, and I'll be damned if I am ever going to stop until I get completely senile. Because I have to, I've got no choice." "Don't let anything stop you," he instructs Zack finally. Toward the end of this scene, Richard Vevers assumes the narrator role: "Losing the Barrier Reef has actually got to mean something," he urges. "Us losing the Great Barrier Reef has got to wake up the world."

Chasing Coral sends out a clear and somber message about corals and climate change. The filmmakers were not interested in offering solutions, Gates explained to me later, but in documenting the depth and extent of the unfolding disaster, thereby making coral lives matter. At the peak of its gloominess, and only four minutes before the credits roll in, the movie offers some relief. Zack, Veron, and Vevers dive in one of the only remaining healthy reefs in the Great Barrier Reef and emerge from it smiling. Next, Zack teaches a group of children about reefs as they gaze into funny-looking boxes containing three-dimensional reef imagery. "It's not too late for coral reefs," Hoegh-Guldberg interjects. Vevers has the last words in the film: "This is inevitable, this great transformation. And that's what makes me so optimistic. All we've got to do is give it a bit of a shove."

Although the film has come to an end, the story it documents continues to unfold and reverberate. The healthy reef shown in the final scene has, in the meantime, died, and the federal scientists whose testimony helped the filmmakers document the extent of bleaching must now navigate NOAA's new, and openly hostile, climate change policies. It turns out that the film has also deepened the divide between coral scientists who were involved in its making and those who were not. Many were finally not at all happy with the film's exclusive focus on climate change.

Clearly, changing our current trajectory will require more than "a bit of a shove." As the coral scientists I interviewed for this project have taught me, staying with coral trouble necessitates a persistent and nuanced being and acting. Following these scientists at their darkest hour and witnessing their insistence to hold on tight to the most precious truths they know—their love of corals and their ultimate belief in the power of life—has made me appreciate such subtleties. The journey from ephemeral optimism that doesn't sustain, to deeper iterations of hope that emerge on the brink of despair and exasperation, requires a leap of faith. In the midst of their enormous pain and loss, coral scientists emerge to guide conservationists, and us all, through the perils of the Anthropocene.

NOTES

INTRODUCTION. CORAL WHISPERERS

1. Les Kaufman (professor of biology, Boston University), interview by author, in person, Boston University, April 5, 2017. Kaufman is part of a collaboration, called the Coral Whisperer, between Boston University and the Virginia-based environmental group Conservation International. See, e.g., "Sustainability at BU," posted November 17, 2014. http://www.bu.edu/sustainability/the-coral-whisperer/. Note: I will be using my initials (IB) and those of the person interviewed (XY) in the interchapter interviews and in a couple of other places throughout the book.

2. NOAA, "Global Coral Bleaching Event Likely Ending," posted June 19, 2017. http://www.noaa.gov/media-release/global-coral-bleaching-event-likely-ending.

3. Will Steffen, Paul J. Crutzen, and John R. McNeill, "The Anthropocene: Are Humans Now Overwhelming the Great Forces of Nature?" *MBIO: A Journal of the Human Environment* 36 (2007): 614–621.

4. Ove Hoegh-Guldberg et al., "Coral Reefs under Rapid Climate Change and Ocean Acidification," *Science* 318 (2007): 1737–1742.

5. Ove Hoegh-Guldberg (director, Global Change Institute; professor of marine science, University of Queensland, Brisbane, Australia), interview by author, Skype, February 25, 2015.

6. *The Guardian*, "Coral Bleaching Has Changed the Great Barrier Reef Forever—Video," June 6, 2016. https://www.theguardian.com/environment/video/2016/jun/07/coral-bleaching-has-changed-the-great-barrier-reef-forever-video.

7. Richard Schiffman, "A Close-Up Look at the Catastrophic Bleaching of the Great Barrier Reef: Interview with Terry Hughes," *Yale Environment 360*, April 10, 2017. https://e360.yale.edu/features/inside-look-at-catastrophic-bleaching-of-the-great-barrier-reef-2017-hughes.

8. Sarah Frias-Torres (marine ecologist, Smithsonian Marine Station, Fort Pierce, Florida), Q&A Symposium, International Coral Reef Symposium, Oahu, HI, June 22, 2016.

9. Joshua E. Cinner et al., "Bright Spots among the World's Coral Reefs," *Nature* 535 (2016): 416–419. See also Ed Yong, "Why Some Coral Reefs Are Thriving: Not All of the World's Reefs Are in Bad Shape—and a Few of the Healthiest Are Managed by Humans," *The Atlantic*, June 15, 2016. http://www.theatlantic.com/science /archive/2016/06/the-surprising-bright-spots-among-the-worlds-coral-reefs/487118/.

10. Madeleine J. H. van Oppen et al., "Building Coral Reef Resilience through Assisted Evolution," *Proceedings of the National Academy of Sciences* 112 (2015): 2307–2313.

11. Smithsonian's Ocean Portal, "From Despair to Repair: Protecting Parrotfish Can Help Bring Back Caribbean Coral Reefs," July 2, 2014. http://ocean.si.edu/blog /despair-repair-protecting-parrotfish-can-help-bring-back-caribbean-coral-reefs. See also Scripps Institution of Oceanography, "Major Multi-Institutional Report Led by Scripps Professor Emeritus Jeremy Jackson Details the Plight and Hope of Caribbean Corals," July 2, 2014. https://scripps.ucsd.edu/news/despair-repair-protecting-parrotfish-can-help-bring-back-caribbean-coral-reefs.

12. Nancy Knowlton, "Why I Am an Ocean Optimist," *The Huffington Post*, June 2, 2015. http://www.huffingtonpost.com/nancy-knowlton/why-i-am-an-ocean-optimist_b_7487286.html. See also Kirsty Nash, "#ICRS2016: Are Coral Reef Scientists All Doom and Gloom?" Kirsty L. Nash, 2016. http://www.kirstynash.com /icrs2016-summary.html.

13. Nancy Knowlton and Jeremy Jackson, "Beyond the Obituaries," *Solutions* 2 (2011): 1. https://www.thesolutionsjournal.com/article/beyond-the-obituaries/.

14. Peter Sale (marine ecologist, professor emeritus, University of Windsor, Canada), interview by author, Skype, July 7, 2016.

15. Elin Kelsey, "The Rise of Ocean Optimism," *Smithsonian Magazine*, June 8, 2016. http://www.smithsonianmag.com/science-nature/rise-ocean-optimism-180959290/.

16. I would like to thank Peter Sale for clarifying this point.

17. See, e.g., Emma Marris, *Rambunctious Garden: Saving Nature in a Post-Wild World* (Bloomsbury Press, 2013); Peter Kareiva, Robert Lalasz, and Michelle Marvier, "Conservation in the Anthropocene," *Breakthrough Journal* 2 (2011): 26–36. In chapter 3, I will provide a more detailed description of coral gardening approaches.

18. See, e.g., Eben Kirksey, *Emergent Ecologies* (Duke University Press, 2015), and Chris D. Thomas, *Inheritors of the Earth: How Nature Is Thriving in an Age of Extinction* (Penguin, 2017).

19. See, e.g., Donna J. Haraway, *Primate Visions: Gender, Race, and Nature in the World of Modern Science* (Psychology Press, 1989).

20. Hoegh-Guldberg, interview, Skype, May 22, 2017.

21. Terry Hughes et al., "Coral Reefs in the Anthropocene," *Nature* 546 (2017): 82–90, at 86.

22. Several scholars have focused their criticism on what they argue is the misleading application of this term to all of *Homo sapiens*, as if we were all equally responsible;

instead, they blame capitalism, global economy—and, particularly, the rich and privileged—for this mess. "A significant chunk of humanity is not party to fossil fuel at all: hundreds of millions rely on charcoal, firewood or organic waste such as dung. . . . [Hence,] species-thinking on climate change is conducive to mystification and political paralysis. It cannot serve as a basis for challenging the vested interests of business-as-usual." Andreas Malm and Alf Hornborg, "The Geology of Mankind? A Critique of the Anthropocene Narrative," *Anthropocene Review* 1 (2014): 62–69, at 65 and 67. See also Jason W. Moore, "The Capitalocene, Part I: On the Nature and Origins of Our Ecological Crisis," *The Journal of Peasant Studies* 44 (2017): 594–630; Donna J. Haraway, *Staying with the Trouble: Making Kin in the Chthulucene* (Duke University Press, 2016), 99–103. Postcolonial historian Dipesh Chakrabarty disagrees with these criticisms, arguing against the reduction of climate change to capitalism: "Unlike in the crises of capitalism, there are no lifeboats here for the rich and the privileged." Dipesh Chakrabarty, "The Climate of History: Four Theses," *Critical Inquiry* 35 (2009): 197–222, at 221.

23. John P. Rafferty, "Anthropocene Epoch," *Encyclopædia Britannica*, August 31, 2016. https://www.britannica.com/science/Anthropocene-Epoch. See also Paul J. Crutzen, "The 'Anthropocene,'" in *Earth System Science in the Anthropocene: Emerging Issues and Problems*, edited by Ekart Ehlers and Thomas Kraftt (Springer, 2006), 5–12; Jan Zalasiewicz et al., "The New World of the Anthropocene," *Environmental Science & Technology* 44: 2228–2231.

24. For an elaborate discussion of the debate on how to classify and name the algal symbiont, see James Bowen, *The Coral Reef Era: From Discovery to Decline* (Springer, 2015), 116–117.

25. The coral polyp secretes a basal calcarious exoskeleton through aragonite deposition. Between the gastrodermis (the cells lining the stomach) and the outer ectodermis is the mesogloea, a surface layer of mucus, secreted by gland cells, that covers the ectodermis. Linda L. Blackall, Bryan Wilson, and Madeleine J. H. van Oppen, "Coral—The World's Most Diverse Symbiotic Ecosystem," *Molecular Ecology* 24 (2015): 5331–5347.

26. Paola Furla et al., "The Symbiotic Anthozoan: A Physiological Chimera between Alga and Animal," *Integrative & Comparative Biology* 45 (2005): 595–604.

27. Bowen, *Coral Reef Era*, 84.

28. Furla et al., "The Symbiotic Anthozoan." But see the critique of the term *host* in Haraway, *Staying with the Trouble*, 67. To those of us who find Greek reassuring, here is a breakdown of the taxonomic classifications pertaining to corals: the phylum Cnidaria comprises about ten thousand species in four classes: Anthozoa, Cubozoa, Scyphozoa, and Hydrozoa. Hexacorallia, a subclass of Anthozoa, encompasses six orders: Scleractinia (stony corals), Actiniaria (sea anemones), Zoantharia (colony-forming, anemone-like organisms), Corallimorpharia (skeleton-less organisms who resemble anemones and scleractinians), Antipatharia (black corals), and Ceriantharia (tube-dwelling anemones). There are approximately 1,300 described scleractinian species, or stony corals. Most of the scientists I interviewed for this project study these

stony corals, an order that first appears in the fossil record in the mid-Triassic (i.e., around 240 million years ago). Blackall et al., "Coral," 5331–5332. See also Madeleine J. H. van Oppen et al., "Shifting Paradigms in Restoration of the World's Coral Reefs," *Global Change Biology* 23 (2017): 3437–3449.

29. Stefan Helmreich, "How Like a Reef: Figuring Coral, 1839–2010," accessed July 11, 2017. Quoting from Osha Gray Davidson. http://reefhelmreich.blogspot.de/.

30. Heterotrophy accounts for 0–66 percent of the fixed carbon incorporated into coral skeletons and can meet 15–35 percent of daily metabolic requirements in healthy corals, and up to 100 percent in bleached corals. Apart from this carbon input, feeding is likely important to most scleractinian corals, since nitrogen, phosphorus, and other nutrients cannot be supplied from photosynthesis by the coral's symbiotic algae. Fanny Houlbreque and Christine Ferrier-Pages, "Heterotrophy in Tropical Scleractinian Corals," *Biological Reviews* 84 (2009): 1–17.

31. Blackall et al., "Coral," 5338. Viruses are the most abundant biological agents in the sea and infect all forms of cellular marine life. Despite their abundance, the viral communities associated with marine invertebrates are virtually undescribed, with the exception of a range of viral pathogens. Only a single viral genome associated with coral has been sequenced to date.

32. Mikhail Matz (professor of integrative biology, University of Texas at Austin), interview by author, Skype, May 2, 2016.

33. Natasha Myers, "Conversations on Plant Sensing," *NatureCulture* 3 (2015): 35–66.

34. Matz, interview, May 2, 2016.

35. Mary Alice Coffroth (professor of Geology, Graduate Program in Evolution, Ecology and Behavior, University at Buffalo, SUNY), e-mail communication, July 13, 2017. See also Angela F. Little, Madeleine J. H. van Oppen, and Bette L. Willis, "Flexibility in Algal Endosymbioses Shapes Growth in Reef Corals," *Science* 304 (2004): 1492–1494; Daniel M. Poland and Mary Alice Coffroth, "Trans-Generational Specificity within a Cnidarian–Algal Symbiosis," *Coral Reefs* 36 (2017): 119–129; and Shelby E. McIlroy and Mary Alice Coffroth, "Coral Ontogeny Affects Early Symbiont Acquisition in Laboratory-Reared Recruits," *Coral Reefs* 36 (2017): 927–932. Ruth Gates added to this that 15 percent of coral species acquire their symbionts via the mothers' eggs. According to Gates, these are the hardiest corals with respect to stress. Ruth Gates (director, Hawai'i Institute of Marine Biology; president, International Society for Reef Studies), e-mail communication, July 17, 2017.

36. NOAA, "What Is Coral Bleaching?" March 17, 2016. http://oceanservice.noaa .gov/facts/coral_bleach.html.

37. This is still debated. "So when they bleach, the question is, are they also capable of bringing in symbionts from the environment, or can they only rely on what is already there?" Howie Lasker (professor of geology and director, Graduate Program in Evolution, Ecology and Behavior, University at Buffalo, SUNY), interview by author, in person, Buffalo, NY, April 13, 2015. As I will detail in chapter 5, Ruth Gates and others are experimenting with using "super" symbionts to strengthen the host.

38. See Andrew C. Baker, Peter W. Glynn, and Bernhard Riegl, "Climate Change and Coral Reef Bleaching: An Ecological Assessment of Long-Term Impacts, Recovery Trends and Future Outlook," *Estuarine, Coastal and Shelf Science* 80 (2008): 435–471.

39. Mark Eakin (coordinator, Coral Reef Watch Program, NOAA), e-mail communication, October 20, 2017.

40. Van Oppen, *Shifting Paradigms*, 2.

41. Sponsel writes that the transition "from viewing reefs as mysterious, powerful, resilient, and treacherous to viewing them as fragile things susceptible to being harmed by everyday human activity resembles many of our histories of attitudes toward large animal predators." Alistair Sponsel, "From Threatening to Threatened: How Coral Reefs Became Fragile," draft, 30 (quoted with permission). For a detailed historical account of this transition, see also Bowen, *Coral Reef Era*, part I.

42. According to Michel Foucault, biopolitics is a form of power that emerged in the eighteenth century, in which "the ancient right to take life or let live was replaced by a power to foster life or disallow it to the point of death." Michel Foucault, *The History of Sexuality* (Vintage Books, 1990), 138.

43. Foucault's body of work on biopolitics has been incredibly productive for the critical study of conservation, interrogating conservation's "foundational goal of affirmatively saving life." Irus Braverman, *Wild Life: The Institution of Nature* (Stanford University Press, 2015), 227. "Even among those species who are deemed threatened," I added in *Wild Life*, "categories and criteria prioritize the ones who are perceived to be *the most threatened of all*: those whose lives are even more, and finally most, worth saving" (229). See also Irus Braverman, "Anticipating Endangerment: The Biopolitics of Threatened Species Lists," *BioSocieties* 12 (2017): 132–157. It is also helpful to recognize that "conservation science is not a homogeneous bloc but is itself replete with debates about which lives must be fostered and who or what is killable, and why. A biopolitical approach to conservation, therefore, contributes to a broader formulation of environmental politics, shifting the focus beyond the conventional cast of global conservation actors . . . to foreground the nitty-gritty, everyday scientific assumptions, discourses, and practices that make nature governable." Christine Biermann and Robert M. Anderson, "Conservation, Biopolitics, and the Governance of Life and Death," *Geography Compass* 11 (2017): e12329, 2.

44. NOAA, "Importance of Coral Reefs," March 25, 2008. http://oceanservice.noaa.gov/education/kits/corals/coral07_importance.html.

45. Isabelle M. Côté and John D. Reynolds (eds.), *Coral Reef Conservation*, vol. 13 (Cambridge University Press, 2006).

46. The Ocean Agency, "The 3rd Global Coral Bleaching Event—2014/2017," accessed June 30, 2017. http://www.globalcoralbleaching.org/.

47. NOAA, "How Do Coral Reefs Protect Lives and Property?" accessed July 3, 2017. http://oceanservice.noaa.gov/facts/coral_protect.html.

48. John E.N. ["Charlie"] Veron, *A Reef in Time: The Great Barrier Reef from Beginning to End* (Harvard University Press, 2010).

49. All information in this paragraph is from Peter Sale, written comments, August 7, 2017.

50. "Climate change is seductive to organizations that want to be taken seriously. Besides being a ready-made meme, it's usefully imponderable. . . . Climate change is everyone's fault—in other words, no one's. We can all feel good about deploring it. . . . To prevent extinctions in the future, it's not enough to curb our carbon emissions. . . . We need to combat the extinctions that are threatened in the present, work to reduce the many hazards that are decimating North American bird populations, and invest in large-scale, intelligently conceived conservation efforts, particularly those designed to allow for climate change. . . . But it only makes sense *not* to do them if the problem of global warming demands the full resources of every single nature-loving group." Excerpts from Jonathan Franzen, "Carbon Capture: Has Climate Change Made It Harder for People to Care about Conservation?" *The New Yorker*, April 26, 2016. https://www.newyorker.com/magazine/2015/04/06/carbon-capture.

51. Jeremy Jackson, "From Despair to Repair 1080," vimeo video, uploaded by Sandy Cannon-Brown, posted July 1, 2014. https://vimeo.com/99653458 (transcribed by author).

52. Dipesh Chakrabarty, "The Politics of Climate Change Is More Than the Politics of Capitalism," *Theory, Culture & Society* 34(2–3) (2017): 25–37, at 29.

53. Claude Lévi-Strauss, *The Savage Mind [La Pensée Sauvage]*, translated by Edmund Leach (University of Chicago Press, 1962).

54. Anthropologist Eva Hayward asks in this regard: "How did sex-assumptions drawn from intuitions about human and mammalian reproduction possibly enriddle corals with anthropocentric and even Euro-American-centric intentionality?" Eva Hayward, "Fingereyes: Impressions of Cup Corals," *Cultural Anthropology* 25 (2010): 578–599, at 588.

55. See, e.g., David Bresnan, "The Birth of the Dolomites: Beautiful Mountains Born out of the Sea," *Scientific American*, June 13, 2012.

56. Toni Makani Gregg et al., "Puka Mai He Koʻa: The Significance of Corals in Hawaiian Culture," in *Ethnobiology of Corals and Coral Reefs*, edited by Nemer Narchi and Lisa Price (Springer, 2015), 2. This also draws on my interview with Misaki Takabayashi (interim associate dean, College of Arts and Sciences, and professor of marine science, University at Hawaiʻi, Hilo), in person, Hilo, HI, June 28, 2016.

57. Karl Marx, *Capital, vol. 1*, chapter 14, section 5, fn. 41 (Online Library of Liberty, 1909; originally published in 1867).

58. Charles Darwin, *Geological Observations on South America: Being the Third Part of the Geology of the Voyage of the Beagle, under the Command of Capt. Fitzroy, RN during the Years 1832 to 1836* (Createspace Independent Publishing Platform; originally published in 1846).

59. Stephen Jay Gould, *The Structure of Evolutionary Theory* (Harvard University Press, 2002), 18. See also Florian Maderspacher, "The Captivating Coral—the Origins of Early Evolutionary Imagery," *Current Biology* 16 (2006): R476–R478.

60. Eugene Rosenberg and Ilana Zilber-Rosenberg, *The Hologenome Concept: Human, Animal and Plant Microbiota* (Springer, 2013), 82. See also Ilana Zilber-

Rosenberg and Eugene Rosenberg, "Role of Microorganisms in the Evolution of Animals and Plants: The Hologenome Theory of Evolution," *FEMS Microbiology Reviews* 32 (2008): 723–735.

61. Gilles Deleuze and Félix Guattari, *A Thousand Plateaus: Capitalism and Schizophrenia* (University of Minnesota Press, 1987; originally published in 1980).

62. Maureen A. O'Malley, William Martin, and John Dupré, "The Tree of Life: Introduction to an Evolutionary Debate," *Biology and Philosophy* 25 (2010): 441–453.

63. But mesophotic reefs at the depth of 30–150 meters do photosynthesize, and they use extraordinary techniques, such as production of fluorescent light for their symbionts, in order to sustain photosynthesis in deeper symbiont layers. See Edward G. Smith et al., "Acclimatization of Symbiotic Corals to Mesophotic Light Environments through Wavelength Transformation by Fluorescent Protein Pigments," *Proceedings of the Royal Society B* 284 (2017): 20170320.

64. J Murray Roberts (professor of marine biology, University of Edinburgh), interview by author, Skype, January 28, 2016.

65. Anna L. Tsing, *The Mushroom at the End of the World: On the Possibility of Life in Capitalist Ruins* (Princeton University Press, 2015).

66. Paul Rabinow, *Making PCR: A Story of Biotechnology* (University of Chicago Press, 2011).

67. Dale Jamieson, *Reason in a Dark Time: Why the Struggle against Climate Change Failed—and What It Means for Our Future* (Oxford University Press, 2014).

68. Joshua P. Howe, *Behind the Curve: Science and the Politics of Global Warming* (University of Washington Press, 2014).

69. James Hansen, *Storms of My Grandchildren: The Truth about the Coming Climate Catastrophe and Our Last Chance to Save Humanity* (Bloomsbury, 2010).

70. Stephen M. Meyer, *The End of the Wild* (MIT Press, 2006).

71. Elizabeth Kolbert, *The Sixth Extinction: An Unnatural History* (Henry Holt, 2014).

72. Bill McKibben, *The End of Nature* (Random House, 1989).

73. Mary Pipher, *The Green Boat: Reviving Ourselves in Our Capsized Culture* (Riverhead Books, 2013).

74. Jonathan Lear, *Radical Hope: Ethics in the Face of Cultural Devastation* (Harvard University Press, 2006).

75. Teresa Shewry, *Hope at Sea: Possible Ecologies in Oceanic Literature* (University of Minnesota Press, 2015).

76. Joanna Macy and Chris Johnstone, *Active Hope: How to Face the Mess We're in without Going Crazy* (New World Library, 2012).

77. Andrew Balmford, *Wild Hope: On the Front Lines of Conservation Success* (University of Chicago Press, 2012).

78. Michael Bloomberg and Carl Pope, *Climate of Hope: How Cities, Businesses, and Citizens Can Save the Planet* (St. Martin's Press, 2017).

79. Ursula K. Heise, *Imagining Extinction: The Cultural Meaning of Endangered Species* (University of Chicago Press, 2016), 12.

80. Haraway, *Staying with the Trouble*, 4.

81. Shewry, *Hope at Sea*, 13–14.

82. Howe, *Behind the Curve*, 7, 9.

83. Jamieson, *Reason in a Dark Time*, 4.

84. Jessica Barnes and Michael R. Dove, "Introduction," in *Climate Cultures: Anthropological Perspectives on Climate Change*, edited by Jessica Barnes and Michael R. Dove (Yale University Press, 2015), 1–21, at 4.

85. Pipher, *Green Boat*, 40–41.

86. Geographers Paul Robbins and Sarah Moore highlight the angst of scientists in the Anthropocene toward normative or political judgments: "Anthropocene scientific culture thus simultaneously displays a panicked political imperative to intervene more vocally and aggressively in an earth transformation run amok and an increasing fear that past scientific claims about the character of ecosystems and their transformation were overly normative, prescriptive, or political in nature. Agonizing over the role of advocacy, especially in conservation, has therefore become a literature in the field all its own." Paul Robbins and Sarah A. Moore, "Ecological Anxiety Disorder: Diagnosing the Politics of the Anthropocene," *Cultural Geographies* 20 (2013): 3–19, at 9.

87. See, in this regard, Sy Montgomery, *Walking with the Apes: Jane Goodall, Dian Fossey, Birute Galdikas* (Houghton Mifflin, 1991). As Jeremy Jackson told me in the context of coral scientists: "All the [International Coral Reef Society] Darwin Medal awardees are old white guys chosen (ironically) by committees with loads of women as well as men. At least four pioneering women coral scientists have been overlooked: Barbara Brown, Nancy Knowlton, Carden Wallace, and Bettie Willis." At the same time, Jackson also notes that "there are a lot of older white men who have been fighting very hard for a more balanced and positive view and are being slandered for it—Stuart Sandin, Forest Rohwer, Mark Vermeij and lots of [other male scientists]." Jeremy Jackson (professor at the Scripps Institution of Oceanography and senior scientist emeritus at the Smithsonian Tropical Research Institute in the Republic of Panama), e-mail communication, October 21, 2017.

88. Although Budke didn't make this explicit, I interpreted this act as one of compassion and care. Both breaking off the corals and placing them back in the water are constitutive of what I have referred to elsewhere as the "biopolitics of conservation," which highlights the entanglements of making live and letting die (Braverman, *Wild Life*). Similarly, Krithika Srinivasan draws on Foucauldian biopolitics and biopower to remind us that "even though biopower is directed at fostering life, violence and harm do not disappear; rather, they are rationalized as necessary for the flourishing of the population. . . . In other words, an entanglement of harm and care goes with the sacrificial logic of population: individuals can be harmfully intervened on in the name of universal wellbeing." Krithika Srinivasan, "Caring for the Collective: Biopower and Agential Subjectification in Wildlife Conservation," *Environment and Planning D: Society and Space* 32 (2014): 501–517, at 508.

89. To convey a sense of the corals' agency, I will use the pronoun *herself*, rather than the neutral *its*, when referring to singular corals (and other living beings).

90. Kaufman, interview, April 5, 2017.

91. Anonymous comment to author, June 20, 2016.

92. Richard J. Lazarus, "Super Wicked Problems and Climate Change: Restraining the Present to Liberate the Future," *Cornell Law Review* 94 (2009): 1153–1233, at 1159.

93. Hoegh-Guldberg et al., "Coral Reefs under Rapid Climate Change."

94. Brooks Barnes, "At Sundance, the Theme Is Climate Change," *The New York Times*, January 10, 2017. https://www.nytimes.com/2017/01/10/movies/at-sundance-the-theme-is-climate-change.html.

INTERVIEW. PETER SALE

1. Peter Sale, *Our Dying Planet: An Ecologist's View of the Crisis We Face* (University of California Press, 2011).

CHAPTER ONE. CORAL SCIENTISTS BETWEEN HOPE AND DESPAIR

1. Mary Pipher, *The Green Boat: Reviving Ourselves in Our Capsized Culture* (Riverhead Books, 2013), 35.

2. Pipher, *Green Boat*, 2.

3. Will Steffen et al., "The Trajectory of the Anthropocene: The Great Acceleration," *The Anthropocene Review* 2 (2015): 81–98. See also Will Steffen, Paul J. Crutzen, and John R. McNeill, "The Anthropocene: Are Humans Now Overwhelming the Great Forces of Nature?" *MBIO: A Journal of the Human Environment* 36 (2007): 614–621; Will Steffen et al., *Global Change and the Earth System: A Planet under Pressure* (Springer, 2005).

4. Donna J. Haraway, *Staying with the Trouble: Making Kin in the Chthulucene* (Duke University Press, 2016), 4.

5. Joanna Macy and Chris Johnstone, *Active Hope: How to Face the Mess We're in without Going Crazy* (New World Library, 2012), 37.

6. Teresa Shewry, *Hope at Sea: Possible Ecologies in Oceanic Literature* (University of Minnesota Press, 2015), 5.

7. See also Ben Halpern (professor of marine ecology, University of California, Santa Barbara), interview by author, in person, International Coral Reef Symposium, Oahu, HI, June 23, 2016.

8. From "Enough Words?" in *The Essential Rumi*, translated by Coleman Barks with John Moyne (Harper, 1995), 20.

9. More recently, scientists have been studying the potential relationship between recent ocean warming and outbreaks of white-band disease. The results have confirmed a strong correlation between the two. Carly J. Randall and Robert Van Woesik, "Contemporary White-Band Disease in Caribbean Corals Driven by Climate Change," *Nature Climate Change* 5 (2015): 375–379.

10. Nilda Jiménez-Marrero (biologist, Puerto Rico Department of Natural Resources), interview by author, in person, Boquerón, PR, January 22, 2014.

11. @SeaCitizens, "Bleaching Is Bad but Giving Up Is Worse #OceanOptimism #EarthOptimism #ICRS2016," Twitter, June 22, 2016, 1:11 P.M. https://twitter.com /SeaCitizens/status/745665511040552960.

12. Ove Hoegh-Guldberg, "It's Not 'Doom and Gloom' to Point Out What's Really Happening to Coral Reefs," *The Conversation*, March 3, 2016. https://theconversation .com/its-not-doom-and-gloom-to-point-out-whats-really-happening-to-coral-reefs-55695.

13. But see postcolonial historian Dipesh Chakrabarty's claim that the focus on emissions is only the first, and a narrow, way of approaching climate change. "Another way to view climate change [is] as part of a complex family of interconnected problems, all adding up to the larger issue of a growing human footprint on the planet that has, over the last couple of centuries and especially since the end of the Second World War, seen a definite ecological overshoot on the part of humanity." Dipesh Chakrabarty, "The Politics of Climate Change Is More Than the Politics of Capitalism," *Theory, Culture & Society* 34(2–3) (2017): 25–37, at 29.

14. Unless indicated otherwise, all the quotes in this section are from Sarah Frias-Torres (marine ecologist, Smithsonian Marine Station, Fort Pierce, Florida), interview by author, in person, International Coral Reef Symposium, Oahu, HI, June 23, 2016. See also Nicole Helgason, "In the Seychelles, Scientists Are Getting Help from Nature's Cleanup Crew," *Reef Builders*, November 9, 2015. https://reefbuilders .com/2015/11/09/seychelles-scientists-natures-cleanup-crew/.

15. Author's observation at the International Coral Reef Symposium, Oahu, HI, June 22, 2016.

16. Frias-Torres, interview, June 23, 2016.

17. Dennis Normile, "Massive Bleaching Killed 35 Percent of the Coral on the Northern End of the Great Barrier Reef," *Science News*, May 30, 2016. http://www .sciencemag.org/news/2016/05/massive-bleaching-killed-35-coral-northern-end-great-barrier-reef.

18. NOAA, "US Coral Reefs Facing Warming Waters, Increased Bleaching," June 20, 2016. http://www.noaa.gov/us-coral-reefs-facing-warming-waters-increased-bleaching.

19. Ibid. See also Ben Rosen, "Even as Bleaching Continues, Hope Remains for Coral Reefs," *The Christian Science Monitor*, June 21, 2016. http://www.csmonitor.com /Science/2016/0621/Even-as-bleaching-continues-hope-remains-for-coral-reefs.

20. Joanie Kleypas (marine ecologist and geologist, National Center for Atmospheric Research, Climate and Global Dynamics), interview by author, in person, International Coral Reef Symposium, Oahu, HI, June 23, 2016.

21. NOAA, "Warm Forecast for Coral Reefs," accessed November 19, 2017. https:// sos.noaa.gov/datasets/warm-forecast-for-coral-reefs/.

22. Ruben Hooidonk et al., "Opposite Latitudinal Gradients in Projected Ocean Acidification and Bleaching Impacts on Coral Reefs," *Global Change Biology* 20 (2014): 103–112. http://onlinelibrary.wiley.com/doi/10.1111/gcb.12394/abstract.

23. NOAA, "Technology and the Study of Coral Reefs: How a Handful of Students Changed the Way NOAA Does Business," revised May 12, 2017. http://celebrating 200years.noaa.gov/magazine/coral_tech/.

24. Coral Reef Watch, NOAA, "NOAA Coral Reef Watch Satellite Bleaching Alert System," accessed on June 30, 2017. http://coralreefwatch-satops.noaa.gov/VS /virtual_station_alerts.html.

25. Ocean Optimism, accessed on March 24, 2018. http://www.oceanoptimism.org/.

26. Nancy Knowlton (Sant Chair in Marine Science, Smithsonian National Museum of Natural History), interview by author, Skype, July 11, 2016.

27. William Skirving (NOAA Coral Reef Watch, Kirwan, Queensland, Australia), interview by author, in person, NOAA Workshop, Hilo, HI, June 28, 2016.

28. Quoted in Daniel Cressey, "Coral Crisis: Great Barrier Reef Bleaching Is 'the Worst We've Ever Seen,'" *Nature*, April 13, 2016. http://www.nature.com/news /coral-crisis-great-barrier-reef-bleaching-isthe-worst-we-ve-ever-seen-1.19747.

29. Graham Readfearn, "Australia's Censorship of UNESCO Climate Report Is Like a Shakespearean Tragedy," *The Guardian*, May 30, 2016. https://www.theguardian .com/environment/planet-oz/2016/may/30/australias-censorship-of-unesco-climate-report-is-like-a-shakespearean-tragedy.

30. Ove Hoegh-Guldberg (director of Global Change Institute and professor of marine science, University of Queensland, Brisbane, Australia), interview by author, Skype, February 25, 2015. A longer part of this interview, and the follow-up conducted almost two years later, immediately follows this chapter.

31. Ibid.

32. Ruth Gates (director, Hawai'i Institute of Marine Biology; president, International Society for Reef Studies), interview by author, Skype, January 25, 2016.

33. Ibid.; see also chapter 5.

34. Gates, interview, January 25, 2016.

35. Ibid. For a detailed account, see Gates's interchapter interview toward the end of this book.

36. Australian Government and Australian Institute of Marine Science, "'Assisted Evolution' versus 'Genetic Modification,'" accessed June 13, 2017. http://www.aims.gov .au/docs/media/featured-content.html/-/asset_publisher/Ydk18I5jDwF7/content /assisted-evolution-versus-genetic-modification.

37. As detailed in the interchapter interview with Gates.

38. Gates, interview, January 25, 2016.

39. Gates, interview, in person, Coconut Island, Oahu, HI, July 1, 2016.

40. "About Coral-List," accessed March 24, 2018. http://coral.aoml.noaa.gov /mailman/listinfo/coral-list.

41. Dennis Hubbard. "Re: Is Any Reef 'Remote?'" Coral-List, May 18, 2017 (cited with permission).

42. Elin Kelsey (associate faculty, School of Environment and Sustainability, Royal Roads University), interview by author, Skype, July 22, 2016.

43. Ibid.

44. Kelsey, interview, in person, Munich, Germany, July 3, 2017.

45. Knowlton, interview, July 11, 2016.

46. Kelsey, interview, July 22, 2016.

47. Ibid.

48. Anonymous interview by author.

49. See, e.g., Johnny Langenheim, "Are Local Efforts to Save Coral Reefs Bound to Fail?" *The Guardian*, August 2, 2016. https://www.theguardian.com/environment /the-coral-triangle/2016/aug/02/are-local-efforts-to-save-coral-reefs-bound-to-fail.

50. Joshua E. Cinner et al., "Bright Spots among the World's Coral Reefs," *Nature* 535 (2016): 416–419.

51. Ibid.

52. Joshua E. Cinner (professorial research fellow and chief investigator, Australian Research Council Centre of Excellence for Coral Reef Studies), interview by author, in person, Waikiki, HI, June 21, 2016.

53. John Bruno and Abel Valdivia, "Coral Reef Degradation Is Not Correlated with Local Human Population Density," *Scientific Reports* 6 (2016): article 29778.

54. Ibid.

55. Writing about forests in Appalachian Ohio, geographer Becky Mansfield et al. similarly investigate two types of "environmental politics after nature," insisting on conceiving "people and their needs, visions, and actions as internal to what nature is and does." Becky Mansfield et al., "Environmental Politics after Nature: Conflicting Socioecological Futures," *Annals of the Association of American Geographers* 105 (2015): 284–293, at 285.

56. Langenheim, "Local Efforts to Save Coral Reefs."

57. Ibid.

58. Margaret W. Miller (research ecologist, NOAA/National Marine Fisheries Service, Southeast Fisheries Science Center), interview by author, telephone, September 16, 2015.

59. Miller, e-mail communication, July 18, 2016.

60. Collectively, cars and trucks account for nearly one-fifth of *all* U.S. emissions, releasing around twenty-four pounds of carbon dioxide and other global-warming gases for every gallon of gas. "Car Emissions and Global Warming," Union of Concerned Scientists, accessed November 19, 2017. http://www.ucsusa.org/clean-vehicles /car-emissions-and-global-warming#.WdO3l4YpBbU.

61. Richard Vevers (founder and CEO, The Ocean Agency), interview by author, Skype, August 5 and 6, 2016.

62. As further described in chapter 2.

63. The Ocean Agency, "Google Underwater Street View," 2017. http://www .theoceanagency.org/underwaterstreetview/.

64. Vevers, interview, August 5 and 6, 2016.

65. Ibid.

66. Ibid.

67. See also Hoegh-Guldberg, interchapter interview.

68. Macy and Johnstone, *Active Hope*, 38.

69. Unless stated otherwise, the quotes in this section are from my interview with Ben Halpern, June 23, 2016.

70. Benjamin Halpern et al., "A Global Map of Human Impact on Marine Ecosystems," *Science* 319 (2008): 948–952. See also Benjamin Halpern et al., "Spatial and Temporal Changes in Cumulative Human Impacts on the World's Ocean," *Nature Communications* 6 (2015): article 7615.

71. The term *heavy impact* means that "fish populations are well below anything close to pristine levels, there's whole groups of species that are missing from it, [and] there's a significant amount of habitat degradation." Halpern, interview, June 23, 2016.

72. National Center for Ecological Analysis and Synthesis, "A Global Map of Human Impacts on Marine Ecosystems," accessed November 28, 2017. https://www.nceas.ucsb.edu/globalmarine.

73. Ocean Health Index, homepage, accessed November 28, 2017. http://www.oceanhealthindex.org/.

74. Lindsay Mosher, "Mapping Human Impacts on the Ocean," Ocean Health Index, August 24, 2015. http://www.oceanhealthindex.org/news/Mapping_Impacts.

75. Ocean Health Index, homepage.

76. Shewry, *Hope at Sea*, 13.

77. *The Essential Rumi*, 50.

INTERVIEW. OVE HOEGH-GULDBERG

1. Ove Hoegh-Guldberg, "Climate Change, Coral Bleaching and the Future of the World's Coral Reefs," *Marine and Freshwater Research* 50 (1999): 839–866.

CHAPTER TWO. "AND THEN WE WEPT"

1. Aldo Leopold, *A Sand County Almanac* (1949; reprint, Ballantine, 1970), 197.

2. NOAA, "NOAA Scientists Report Mass Die Off of Invertebrates at East Flower Garden Bank in Gulf of Mexico," July 2016. http://sanctuaries.noaa.gov/news/jul16/noaa-scientists-report-mass-die-off-of-invertebrates-at-east-flower-garden-bank.html.

3. Ibid.

4. Kristina Gjerde (adjunct professor, Middlebury Institute of International Studies at Monterey; senior high seas advisor, International Union for the Conservation of Nature), interview by author, Skype, June 7, 2016. For a brief discussion of cold-water corals in this book, see the interchapter interview with Roberts.

5. NOAA, "Mass Die Off of Invertebrates at East Flower Garden Bank."

6. See chapter 5 for further details.

7. James Bowen, *The Coral Reef Era: From Discovery to Decline* (Springer, 2015), 154.

8. Angela E. Douglas, "Coral Bleaching—How and Why?" *Marine Pollution Bulletin* 46 (2003): 385–392.

9. For the genealogy of bleaching studies in coral science see Bowen, *Coral Reef Era*, 153–157.

10. Mark Eakin (coordinator, NOAA Coral Reef Watch), e-mail communication, October 20 and November 29, 2017.

11. Lauretta Burke et al., *Reefs at Risk Revisited* (World Resources Institute, 2011).

12. Iain McCalman, *The Reef: A Passionate History: The Great Barrier Reef from Captain Cook to Climate Change* (Macmillan, 2014), 270.

13. Ibid., 271.

14. Mikhail Matz (professor of integrative biology, University of Texas at Austin), interview by author, Skype, May 2, 2016.

15. McCalman, *Reef: A Passionate History*, 281.

16. Eakin, interview, telephone, November 16, 2015.

17. Mary Alice Coffroth (professor of geology, Graduate Program in Evolution, Ecology and Behavior, University at Buffalo, SUNY), interview by author, in person, Buffalo, New York, November 16, 2015.

18. Mary Pipher, *The Green Boat: Reviving Ourselves in Our Capsized Culture* (Riverhead Books, 2013), 20.

19. Ibid., 23, quoting from Joanna Macy, Daniel Goleman, and Glenn Albrecht, respectively. Finally, the term "ecological anxiety disorder" was coined in Paul Robbins and Sarah A. Moore, "Ecological Anxiety Disorder: Diagnosing the Politics of the Anthropocene," *Cultural Geographies* 20 (2013): 3–19.

20. Pipher, *Green Boat*, 20.

21. David R. Bellwood et al., "Confronting the Coral Reef Crisis," *Nature* 429 (2004): 827–833; Hughes et al., "Coral Reefs in the Anthropocene." See also Ove Hoegh-Guldberg, "Climate Change, Coral Bleaching and the Future of the World's Coral Reefs," *Marine and Freshwater Research* 50 (1999): 839–866.

22. Similarly, in *Twice Dead*, Margaret Lock explores the time that elapses between brain and heart deaths to suggest that "the space between life and death is historically and culturally constructed, fluid, multiple, and open to dispute." Margaret Lock, *Twice Dead: Organ Transplants and the Reinvention of Death* (University of California Press, 2001), 11.

23. Rachel Nuwer, "Stressed Corals Dim Then Glow Brightly before They Die," *Smithsonian*, March 13, 2013. http://www.smithsonianmag.com/science-nature/stressed-corals-dim-then-glow-brightly-before-they-die-2197598/#cQsbByaiK2FLboO8.99.

24. Jacqueline Ronson, "What Is Coral Bleaching? How Some Reefs Glow before They Die," *Inverse Science*, July 14, 2017. https://www.inverse.com/article/34218-what-is-coral-bleaching-chasing-coral-great-barrier-reef-dead.

25. Nuwer, "Stressed Corals."

26. Richard Vevers (founder and CEO, The Ocean Agency), interview by author, Skype, August 5 and 6, 2016.

27. Douglas H. Chadwick, "Corals in Peril," *National Geographic* 195 (1999): 30–37, at 37.

28. Irus Braverman, "*Bleached!* Managing Coral Catastrophe," *Futures* 92 (2016): 12–28.

29. Sylvia Earle, *Sea Change: A Message of the Oceans* (Ballantine Books, 1995), xii.

30. Zoë Schlanger, "If Oceans Stopped Absorbing Heat from Climate Change, Life on Land Would Average 122°F," *Quartz*, November 29, 2017. https://qz.com/1141633 /if-oceans-stopped-absorbing-heat-from-climate-change-life-on-land-would-average-122f/.

31. Elizabeth Kolbert, *The Sixth Extinction: An Unnatural History* (Henry Holt, 2014), 3.

32. Jeremy Walker, "The Creation to Come: Directing the Evolution of the Bio-economy," in *Environmental Change and the World's Futures: Ecologies, Ontologies, and Mythologies*, edited by Jonathan Paul Marshall and Linda Connor (Routledge, 2015), 264–281.

33. Mark D. Spalding, Corinnea Ravilious, and Edmund P. Green, "United Nations Environment Programme, World Conservation Monitoring Centre," in *World Atlas of Coral Reefs* (University of California Press, 2001).

34. Although his perspective is not shared by the vast majority of conservationists, it is worth mentioning that ecologist Chris D. Thomas offers a decidedly different—and quite hopeful—interpretation of the sixth extinction. He argues that alongside the mass extinction of our time, an equal, if not greater, mass *genesis* is occurring. He bases this on the number of novel and often marginalized species that are thriving in our hybrid environments. See Chris D. Thomas, *Inheritors of the Earth: How Nature Is Thriving in an Age of Extinction* (Penguin Random House, 2017).

35. Michael E. Soulé, "What Is Conservation Biology?" *BioScience* 35 (1985): 727–734. In the 1990s, conservation biologists such as Daniel F. Doak and L. Scott Mills called for a move away from what has become, in their view, the "crisis discipline" of conservation biology, toward a complete reevaluation of the field. Daniel F. Doak and L. Scott Mills, "A Useful Role for Theory in Conservation," *Ecology* 75 (1994): 615–626.

36. For a discussion of the use of algorithms in conservation science, see Irus Braverman, "Governing the Wild: Databases, Algorithms, and Population Models as Biopolitics," *Surveillance & Society* 12 (2014): 15–37.

37. NOAA, "Celebrating 200 Years," accessed March 23, 2018. http://celebrating 200years.noaa.gov/magazine/coral_tech/.

38. Stefan Helmreich, *Alien Ocean: Anthropological Voyages in Microbial Seas* (University of California Press, 2009), 25.

39. NOAA, "Celebrating 200 Years."

40. Eakin, interview, November 16, 2015.

41. Helmreich, *Alien Ocean*, 38.

42. Ben Halpern (professor of marine ecology, University of California Santa Barbara), interview by author, in person, International Coral Reef Symposium, Oahu, HI. June 23, 2016.

43. Patrick O'Malley, "The Uncertain Promise of Risk," *Australian and New Zealand Journal of Criminology* 37 (2004): 323–343, quoted in Claudia Aradau and Rens van Munster, *The Politics of Catastrophe* (Routledge, 2011), 5.

44. Eakin, interview, November 16, 2015.

45. Ibid.

46. NOAA, "Coral Reef Watch (CRW) Satellite Bleaching Alert (SBA) System," accessed June 30, 2017. http://coralreefwatch-satops.noaa.gov/VS/virtual_station_alerts.html.

47. Ibid. Another technology developed by Coral Reef Watch is the "operational product suite." See NOAA, "Coral Reef Watch Satellite Monitoring," accessed June 30, 2017. http://coralreefwatch.noaa.gov/satellite/index.php.

48. Eakin, interview, November 16, 2015.

49. NOAA, "Daily 5km Satellite Coral Bleaching Heat Stress Monitoring," accessed June 30, 2017. http://coralreefwatch.noaa.gov/satellite/bleaching5km/index.php.

50. Gang Liu et al., "Reef-Scale Thermal Stress Monitoring of Coral Ecosystems: New 5-km Global Products from NOAA Coral Reef Watch," *Remote Sensing* 6 (2014): 11579–11606, at 11585.

51. Eakin, interview, November 16, 2015.

52. Ibid.

53. Helmreich, *Alien Ocean*, 41 (emphasis in original). See also Geoffrey C. Bowker, "Biodiversity Datadiversity," *Social Studies of Science* 30 (2000): 643–683; Rafi Youatt, "Counting Species: Biopower and the Global Biodiversity Census," *Environmental Values* 17 (2008): 393–417.

54. Chandra Mukerji, *A Fragile Power: Scientists and the State* (Princeton University Press, 1990), 153.

55. Eakin, interview, November 16, 2015.

56. Aradau and van Munster, *Politics of Catastrophe*, 113.

57. Alexandra Witze, "Marine Ecologists Take to the Skies to Study Coral Reefs," *Nature News*, May 31, 2016. See also chapter 4 and interchapter interview with Roberts.

58. Witze, "Marine Ecologists Take to the Skies."

59. Ibid.

60. Clinton B. Edwards et al., "Large-Area Imaging Reveals Biologically Driven Non-random Spatial Patterns of Corals at a Remote Reef," *Coral Reefs* 36 (2017): 1291–1305.

61. "100 Island Challenge," accessed November 27, 2017. http://100islandchallenge .org/home/about/. See also Joanna Klein, "Crazy Jigsaw Puzzles Improve Our Views of Coral Reefs," *The New York Times*, November 22, 2017. https://www.nytimes .com/2017/11/22/science/coral-reefs-3d-photomosaics.html.

62. Daniel Cressey, "Coral Crisis: Great Barrier Reef Bleaching Is 'The Worst We've Ever Seen,'" *Scientific American*, April 13, 2016. https://www.scientificamerican.com /article/coral-crisis-great-barrier-reef-bleaching-is-the-worst-we-ve-ever-seen/.

63. Dennis Normile, "Massive Bleaching Killed 35% of the Coral on the Northern End of the Great Barrier Reef," *Science*, May 30, 2016.

64. @ProfTerryHughes, "I showed the results of aerial surveys of #bleaching on the #GreatBarrierReef to my students, and then we wept." Twitter, April 19, 2016, 3:48 P.M. https://twitter.com/ProfTerryHughes.

65. The camera captured this change in my posture and approach as I was caught in the frame a few times. See Larissa Rhodes, *Chasing Coral*, directed by Jeff Orlowski (Exposure Labs and Netflix, 2017).

66. Chris Mooney, "'And Then We Wept': Scientists Say 93 Percent of the Great Barrier Reef Now Bleached," *The Washington Post*, April 20, 2016. https://www.washingtonpost.com/news/energy-environment/wp/2016/04/20/and-then-we-wept-scientists-say-93-percent-of-the-great-barrier-reef-now-bleached/?utm_term=.c6b346a61a35. See also Michael Slezak, "The Great Barrier Reef: A Catastrophe Laid Bare," *The Guardian*, June 6, 2016. https://www.theguardian.com/environment/2016/jun/07/the-great-barrier-reef-a-catastrophe-laid-bare; Rowan Jacobsen, "Obituary: Great Barrier Reef (25 Million BC–2016)," *Outside*, October 11, 2016. https://www.outsideonline.com/2112086/obituary-great-barrier-reef-25-million-bc-2016.

67. Terry Hughes, oral presentation, International Coral Reef Symposium, June 21, 2016, Oahu, HI.

68. Ibid.

69. Ibid.

70. Ibid.

71. In Pauly's words, "Essentially, this symptom has arisen because each generation of fisheries scientists accepts as a baseline the stock size and species composition that occurred at the beginning of their careers, and uses this to evaluate changes. When the next generation starts its career, the stocks have further declined, but it is the stocks at that time that serve as a new baseline. The result obviously is a gradual shift of the baseline, a gradual accommodation of the creeping disappearance of resource species, and inappropriate reference points for evaluating economic losses resulting from over-fishing, or for identifying targets for rehabilitation measures." Daniel Pauly, "Anecdotes and the Shifting Baseline Syndrome of Fisheries," *Trends in Ecology & Evolution* 10 (1995): 430.

72. Hughes, oral presentation, June 21, 2016.

73. Ibid.

74. For a more detailed definition of the shifting baseline syndrome, see note 71.

75. Dennis Hubbard, Coral-List, May 15, 2017 (cited with permission).

76. Vevers, interview, August 5 and 6, 2016.

77. Damien Cave and Justin Gillis, "Large Sections of the Great Barrier Reef Are Now Dead, Scientists Find," *The New York Times*, March 15, 2017. https://www.nytimes.com/2017/03/15/science/great-barrier-reef-coral-climate-change-dieoff.html?_r=0.

78. Elena Becatoros, "More Than 90 Percent of Coral Reefs Will Die by 2050," *The Independent*, March 13, 2017. http://www.independent.co.uk/environment/environment-90-percent-coral-reefs-die-2050-climate-change-bleaching-pollution-a7626911.html.

79. However, the assumptions behind these strict divides between the geological and the biological as respectively alive and dead are not as straightforward as they may seem. According to philosopher Monika Bakke, mineralization (or calcification), which leads to the formation of stromatolites, coral reefs, shells, bones, and egg shells, is a trace of the affinity between biological and mineral species. She quotes from Manuel DeLanda: "The human endoskeleton was one of the many products of the ancient mineralization. Yet that is not the only geological infiltration that the human species has undergone. About eight thousand years ago, human populations began mineralizing again when they developed an urban exoskeleton." Monika Bakke, "Art and Metabolic Force in Deep Time Environments," *Environmental Philosophy* 14 (2017): 41–59, at 52. See also Kathryn Yusoff, "Geologic Life: Prehistory, Climate, Futures in the Anthropocene," *Environment and Planning D* 31 (2013): 779–795; Nigel Clark and Kathryn Yusoff, "Geosocial Formations and the Anthropocene," *Theory, Culture & Society* 34(2–3) (2017): 3–23.

80. Jacobsen, "Obituary: Great Barrier Reef."

81. Chris D'Angelo, "Great Barrier Reef Obituary Goes Viral, to the Horror of Scientists," *The Huffington Post*, October 14, 2016. http://www.huffingtonpost.com /entry/scientists-take-on-great-barrier-reef-obituary_us_57fff8f1e4b0162c043b068f.

82. Les Kaufman (professor of biology, Boston University), interview by author, in person, Boston, MA, April 5, 2017.

83. Joanna Macy and Chris Johnstone, *Active Hope: How to Face the Mess We're in without Going Crazy* (New World Library, 2012), 3.

84. Hughes, oral presentation, June 21, 2016.

85. Mike Hulme, "Reducing the Future to Climate: A Story of Climate Determinism and Reductionism," *Osiris* 26 (2011): 245–266.

86. See also Elizabeth Kolbert, "Going Negative: Can Carbon-Dioxide Removal Save the World?" *The New Yorker*, November 20, 2017. https://www.newyorker.com /magazine/2017/11/20/can-carbon-dioxide-removal-save-the-world.

87. Dipesh Chakrabarty, "The Politics of Climate Change Is More Than the Politics of Capitalism," *Theory, Culture & Society* 34(2–3) (2017): 25–37, at 26.

88. D'Angelo, "Great Barrier Reef Obituary Goes Viral."

89. Mooney, "And Then We Wept."

90. Ben Anderson discusses preemption, precaution, and preparedness in his writing about the present anticipation of future catastrophe, suggesting that anticipatory action has been formalized and legitimized in response to a number of major threats to liberal-democratic life. The existing threats are depicted as sharing several common characteristics: they are potentially catastrophic—namely, each can irreversibly alter the conditions of life; the source of the disaster is somewhat vague; and the disaster is imminent—i.e., without some form of action, "a threshold will be crossed and a disastrous future will come about." Ben Anderson, "Preemption, Precaution, Preparedness: Anticipatory Action and Future Geographies," *Progress in Human Geography* 34 (2010): 777–798. Counter to Ulrich Beck's thesis regarding the "incalculability" of certain modern risks, Anderson argues that a range of practices have been deployed to render the future present. The first practice he discusses, which is also the most

relevant to coral conservation, is calculation. In this context, Anderson emphasizes the importance of numbers, "which are then visualized in forms of 'mechanical objectivity' such as tables, charts, and graphs." Next, he highlights the extensive use of catastrophe modeling—for example, algorithmic models utilized in the insurance industry to predict and calculate loss by stochastic events (ibid., 783–784). The emerging literature on the present anticipation of futures includes Barbara Adam and Chris Groves, *Future Matters: Action, Knowledge, Ethics* (Koninkijke, Brills, 2007), and Braverman, "Bleached!"

91. NOAA, "Celebrating 200 Years."

92. Ibid.

93. See, e.g., Dahr Jamail, "Coral Reefs Could All Die Off by 2050," *EcoWatch*, May 16, 2017. https://www.ecowatch.com/coral-reef-bleaching-2408656490.html. See also Bob Kearney, "Governments Are Not Protecting the Barrier Reef," *The Conversation*, July 19, 2013. http://theconversation.com/governments-are-not-protecting-the-great-barrier-reef-16107.

94. Achille Mbembe writes that "the notion of biopower is insufficient to account for contemporary forms of subjugation of life to the power of death. . . . I have put forward the notion of necropolitics and necropower to account for the various ways in which, in our contemporary world, weapons are deployed in the interest of maximum destruction of persons and the creation of *death-worlds*, new and unique forms of social existence in which vast populations are subjected to conditions of life conferring upon them the status of *living dead*." Achille Mbembe, "Necropolitics," *Public Culture* 15 (2003): 11–40, at 39–40 (emphasis in original).

95. Eakin, interview, November 16, 2015.

96. Paul Marshall and Heidi Schuttenberg, *A Reef Manager's Guide to Coral Bleaching* (Great Barrier Reef Marine Park Authority, 2006), accessed June 30, 2017. http://www.coris.noaa.gov/activities/reef_managers_guide/reef_managers_guide.pdf.

97. See chapter 1 and interchapter interview with Hoegh-Guldberg.

98. Vevers, interview, August 5 and 6, 2016.

99. Coral-List, March 2017.

100. Eakin, interview, November 16, 2015.

101. Peter Sale (marine ecologist, professor emeritus, University of Windsor, Canada), written comments to author, August 7, 2017.

102. Terry Hughes et al., "Coral Reefs in the Anthropocene," *Nature* 546 (2017): 82–90.

103. Ibid., 86.

104. Ibid., 87–88.

105. Ibid., 88.

INTERVIEW. JEREMY JACKSON

1. For a comprehensive history of tropical biology in these sites (and others), see Jan Sapp, *Coexistence: The Ecology and Evolution of Tropical Biodiversity* (Oxford University Press, 2016), 136–150.

2. Jeremy Jackson and Leo Buss, "Alleopathy and Spatial Competition among Coral Reef Invertebrates," *Proceedings of the National Academy of Sciences* 72 (1975): 5160–5163.

3. Jeremy D. Woodley et al., "Hurricane Allen's Impact on a Jamaican Coral Reef," *Science* 214 (1981): 749–755.

4. Jeremy Jackson et al., "Ecological Effects of a Major Oil Spill on Panamanian Coastal Marine Communities," *Science* 243 (1989): 37–44.

5. Jeremy Jackson, "Reefs since Columbus," *Coral Reefs* 16 (1997): S23–S32.

6. Terry Hughes et al., "Coral Reefs in the Anthropocene," *Nature* 546 (2017): 82–90.

7. Jeremy Jackson et al. (eds.), *Status and Trends of Caribbean Coral Reefs: 1970–2012* (IUCN, 2012). https://cmsdata.iucn.org/downloads/caribbean_coral_reefs___status _report_1970_2012.pdf.

8. Hughes et al., "Coral Reefs in the Anthropocene."

9. The "coral triangle" is the triangular area of the tropical marine waters of Indonesia, Malaysia, Papua New Guinea, Philippines, Solomon Islands, and Timor-Lest.

CHAPTER THREE. FRAGMENTS OF HOPE

1. Scott Russell Sanders, *Hunting for Hope: A Father's Journey* (Beacon Press, 1995), 3.

2. Nilda Jiménez-Marrero (biologist, Puerto Rico Department of Natural Resources), interview by author, telephone, October 7, 2014.

3. See, for example, Irus Braverman, *Zooland: The Institution of Captivity* (Stanford University Press, 2012), and Irus Braverman, "Captive for Life: Conserving Extinct Species through *Ex Situ* Breeding," in *The Ethics of Captivity*, edited by Lori Gruen (Oxford University Press, 2014), 193–212.

4. See chapter 1 and figure 8.

5. For a more detailed description of that event, see chapter 1.

6. Baruch Rinkevich (senior scientist, National Institute of Oceanography, Haifa, Israel), interview by author, telephone, June 18, 2015.

7. See, e.g., Heike K. Lotze, Richard C. Hoffmann, and Jon M. Erlandson, "Lessons From Historical Ecology and Management," *Marine Ecosystem-Based Management* 6 (2014): 17–55.

8. Vitalis Dubininkas, "Effects of Substratum on the Growth and Survivorship of *Montipora capitata* and *Porites lobata* Transplants," *Journal of Experimental Marine Biology and Ecology* 486 (2017): 134–139.

9. See, e.g., Yael B. Horoszowski-Fridman and Baruch Rinkevich, "Restoration of the Animal Forests: Harnessing Silviculture Biodiversity for Coral Transplantation," in *Marine Animal Forests: The Ecology of Benthic Biodiversity Hotspots*, edited by Sergio Rossi et al. (Springer International, 2015), 1–26.

10. Matthew Chrulew, "Managing Love and Death at the Zoo: The Biopolitics of Endangered Species Preservation," *Australian Humanities Review* 50 (2011): 137–157, at 147. See also Elizabeth Hennessy, "Producing 'Prehistoric' Life: Conservation Breed-

ing and the Remaking of Wildlife Genealogies," *Geoforum* 49 (2013): 71–80; and Christine Biermann and Robert M. Anderson, "Conservation, Biopolitics, and the Governance of Life and Death," *Geography Compass* 11 (2017): e12329, 2.

11. As discussed in chapter 4. See also Kirsty Nancarrow and Allyson Horn, "Crown-of-Thorns Starfish: New Method Kills More Than 250,000 Marine Pests on Great Barrier Reef," *ABC News Australia*, April 22, 2014. http://www.abc.net.au/news/2014–04–22/new-method-kills-more-than-2502c000-crown-of-thorns-starfish-0/5403600.

12. Anika Gupta, "Invasion of the Lionfish," *Smithsonian*, May 7, 2009. https://www.smithsonianmag.com/science-nature/invasion-of-the-lionfish-131647135/.

13. Jennifer Billock, "New Priority for Ocean Resorts: Restoring Reefs," *The New York Times*, February 23, 2017. https://www.nytimes.com/2017/02/23/travel/restoring-coral-reefs-ocean-resort-priority.html.

14. Madeleine J. H. van Oppen et al., "Building Coral Reef Resilience through Assisted Evolution," *Proceedings of the National Academy of Sciences* 112 (2017): 2307–2313. See also Madeleine J. H. van Oppen et al., "Shifting Paradigms in Restoration of the World's Coral Reefs," *Global Change Biology* (2017): 3437–3448.

15. Assaf Shaham (coral farmer, OkCoral, Ein Yahav, Israel), interview by author, in person, and observation, Ein Yahav, Israel, June 8, 2015. Anthropologist Stefan Helmreich writes in this context: "Here human kinship with the sea finds a link through the shared substance of hydroxyapatite, the complex calcium-phosphate salt found in human bones and coral skeletons. Future medicine promises to reverse the logic of Shakespeare's famous line from *The Tempest*: 'of his bones are coral made,' as coral becomes an organ donor for humans." Stefan Helmreich, "How Like a Reef: Figuring Coral, 1839–2010," note xxviii, accessed July 12, 2017. http://reefhelmreich.blogspot .de/. Mineralization, or calcification, "which leads to the formation of stromatolites, coral reefs, shells, exoskeletons, bones, and egg shells, but also to kidney and gall stones, is a trace of the affinity between biological and mineral species. Not surprisingly[,] then[,] coral skeletons are highly biocompatible with ours, and are therefore used for human bone transplants and as scaffolding in tissue culture." Monika Bakke, "Art and Metabolic Force in Deep Time Environments," *Environmental Philosophy* 14 (2017): 41–59, at 52.

16. For a discussion of these differences in the terrestrial context, see Irus Braverman, *Wild Life: The Institution of Nature* (Stanford University Press, 2015).

17. Baruch Rinkevich, "Restoration Strategies for Coral Reefs Damaged by Recreational Activities: The Use of Sexual and Asexual Recruit," *Restoration Ecology* 3 (1995): 241–251.

18. Ibid., 241.

19. Craig A. Downs et al., "Toxicopathological Effects of the Sunscreen UV Filter, Oxybenzone (Benzophenone-3), on Coral Planulae and Cultured Primary Cells and Its Environmental Contamination in Hawai'i and the U.S. Virgin Islands," *Archives of Environmental Contamination & Toxicology* 70 (2016): 265–288.

20. Michael Sweet, Andrew Ramsey, and Mark Bulling, "Designer Reefs and Coral Probiotics: Great Concepts but Are They Good Practice?" *Biodiversity* 18 (2017): 19–22,

at 20. See also James R. Pratt, "Artificial Habitats and Ecosystem Restoration: Managing for the Future," *Bulletin of Marine Science* 55 (1994): 268–275.

21. Society for Ecological Restoration, homepage, accessed July 12, 2017. http://www.ser.org/. See also Elisa Bayraktarov et al., "The Cost and Feasibility of Marine Coastal Restoration," *Ecological Applications* 26 (2016): 1055–1074.

22. Rinkevich disagrees, pointing out that calculations of restoration costs are overestimated and that transplant survivorship is underestimated, which leads to an inaccurate evaluation of the cost-benefit ratio of restoration. Baruch Rinkevich, "Rebutting the Inclined Analyses on the Cost-Effectiveness and Feasibility of Coral Reef Restoration," *Ecological Applications* 27 (2017): 1970–1973. In their analysis of the history of recovery in Hawaiian coral reefs, Kittinger et al. found that these reefs have undergone "previously undetected recovery periods . . . including a historical recovery" during the period 1400–1820 and "an ongoing recovery" that started around 1950." John N. Kittinger et al., "Historical Reconstruction Reveals Recovery in Hawaiian Coral Reefs," *PLoS ONE* 6 (2011): e25460. In other words, contrary to popular belief, humans have had restoration-type interactions with coral reef environments before the modern coral restoration movement.

23. Peter Sale (marine ecologist, professor emeritus, University of Windsor, Canada), written comments, August 7, 2017.

24. The Nature Conservancy, "A Scientist Diver Talks about His 'Office' 60 Feet Deep and Why There's Hope for Coral Reefs," accessed July 15, 2017. https://www.nature.org/ourinitiatives/regions/caribbean/underwater-science-dives-deep-into-coral-reef-research.xml.

25. David Mouillot et al., "Global Marine Protected Areas Do Not Secure the Evolutionary History of Tropical Corals and Fishes," *Nature Communications* 7 (2016): article 10359.

26. Baruch Rinkevich, "Management of Coral Reefs: We Have Gone Wrong When Neglecting Active Reef Restoration," *Marine Pollution Bulletin* 56 (2008): 1821–1824.

27. Ibid., 1823.

28. Rinkevich, interview, in person, Waikiki, HI, June 21–22, 2016.

29. Zac H. Forsman (researcher at Hawai'i Institute of Marine Biology and coral specialist for the State of Hawai'i), interview by author, telephone, March 1, 2016.

30. Sale, written comments, August 7, 2017.

31. Sarah Frias-Torres (marine ecologist, Smithsonian Marine Station, Fort Pierce, Florida), interview by author, in person, International Coral Reef Symposium, Oahu, HI, June 23, 2016.

32. Rinkevich, "Management of Coral Reefs," 1822.

33. Thomas J. Goreau and Raymond L. Hayes, "Coral Bleaching and Ocean 'Hot Spots,'" *AMBIO* 23 (1994): 176–180. See also Ariel E. Lugo, Caroline S. Rogers, and Scott W. Nixon, "Hurricanes, Coral Reefs and Rainforests: Resistance, Ruin and Recovery in the Caribbean," *AMBIO* 29 (2000): 106–114.

34. Rinkevich, interview, June 18, 2015.

35. Ibid.

36. Rinkevich, "Management of Coral Reefs," 1822 (in-text citations omitted).

37. Sergio Rossi et al., "Animal Forests of the World: An Overview," in *Marine Animal Forests: The Ecology of Benthic Biodiversity Hotspots*, edited by Sergio Rossi et al. (Springer International, 2015), 1–42.

38. Rinkevich, interview, June 18, 2015.

39. From the Scripps Institution of Oceanography: "Corals are superb fighters. They fend off encroaching corals by stinging them with their tentacles and by ejecting their stomachs to digest them. These coral-coral battle zones are easy to spot; when two different corals meet, there's often a cleared band between the two where they've killed each other off. The same tactics are put to use to fight off invading algae. On healthy reefs, corals can maintain their territory, often beating back and even killing various types of algae. On degraded reefs, the tables are turned. Here the algae are the superior competitors with their own arsenal of weapons including chemical poisons and introducing bacteria that make the coral sick. Sometimes they directly overgrow and smother the coral." Katie Barott, "Battle Zones," Scripps Institution of Oceanography, November 24, 2010. https://scripps.ucsd.edu/labs/coralreefecology/battle-zones/.

40. James E. Maragos, "Coral Transplantation: A Method to Create, Preserve and Manage Coral Reefs," *University of Hawaii Sea Grant Advisory Report* (1974): 35. See also Charles Birkeland, Richard H. Randall, and Gretchen Grimm, *Three Methods of Coral Transplantation for the Purpose of Reestablishing a Coral Community in the Thermal Effluent Area at the Tanguisson Power Plant*, University of Guam Marine Laboratory Technical Report no. 60 (1979).

41. Claude Bouchon, Jean Jaubert, and Y. Bouchon-Navarro, "Evolution of a Semi-artificial Reef Built by Transplanting Coral Heads," *Tethys* 10 (1981): 173–176. See also K. W. Fucik, T. J. Bright, and K. S. Goodman, "Measurements of Damage, Recovery, and Rehabilitation of Coral Reefs Exposed to Oil," in *Restoration of Habitats Impacted by Oil Spills*, edited by John Cairns and Arthur L. Buikema (Butterworth, 1984), 115–133.

42. Rinkevich, "Restoration Strategies," 244.

43. Ibid.

44. Gideon Levy et al., "Mid-Water Rope Nursery—Testing Design and Performance of a Novel Reef Restoration Instrument," *Ecological Engineering* 36 (2010): 560–569. See also Shai Shafir and Baruch Rinkevich, "Integrated Long-Term Mid-Water Coral Nurseries: A Management Instrument Evolving into a Floating Ecosystem," *University of Mauritius Research Journal* 16 (2010): 365–386.

45. Levy et al., "Mid-Water Rope Nursery," 561–566.

46. Ibid.

47. Rinkevich, interview, June 18, 2015.

48. Ibid.

49. Haniya Rae, "For the Success of Coral Restoration, A Matter of Scale," *UnDark*, September 19, 2017. https://undark.org/article/coral-reefs-regrowth-restoration/. See also Nicholas A. J. Graham et al., "Managing Resilience to Reverse Phase Shifts in Coral Reefs," *Frontiers in Ecology and the Environment* 11 (2013): 541–548.

50. Rae, "Success of Coral Restoration."

51. "Our restoration efforts could also help us understand how to enhance the genetic diversity of corals, by tracking all the different genotypes that we have in our coral restoration nurseries and maximizing that diversity when we out-plant." James Byrne, "Marine Restoration Week: The Future of Coral Restoration Science," Nature.org, June 6, 2013. https://blog.nature.org/science/2013/06/06/future-coral-reef-restoration-science/.

52. Rinkevich, interview, telephone, May 1, 2017.

53. Baruch Rinkevich, "Climate Change and Active Reef Restoration: Ways of Constructing the 'Reefs of Tomorrow,'" *Journal of Marine Science and Engineering* 2 (2015): 111–127, at 111 (abstract).

54. Rae, "Success of Coral Restoration."

55. For more details, see interchapter interview with Nedimyer.

56. Walt Smith (founder and director, Walt Smith International Aquaculture, Fiji), interview by author, Skype, July 18, 2016. See also interchapter interview with Nedimyer and accompanying figures.

57. Ken Nedimyer (founder and president, Coral Restoration Foundation), interview by author, telephone, October 17, 2017. For more about the tree nursery, see interchapter interview with Nedimyer. Although such tree-like structures have much better growth rates, their unnatural location in the water column has raised concerns among some coral restoration scientists about introducing selective features to the corals who thrive on them.

58. Rinkevich, interview, May 1, 2017. Becky Mansfield et al. complicate Rinkevich's notion of forests and silviculture, identifying six types of "socioecological forests" in Appalachian Ohio and calling for a social nature that does not assume, a priori, that one is naturally better than another (though, admittedly, they do not include tree nurseries in their forest types). Becky Mansfield et al., "Environmental Politics after Nature: Conflicting Socioecological Futures," *Annals of the Association of American Geographers* 105 (2015): 284–293.

59. Forsman, e-mail communication, July 27, 2016.

60. Forsman, interview, in person, Waikiki, HI, June 20, 2016.

61. *Smithsonian Portal*, "The Coral Gardener," accessed July 5, 2017. http://ocean.si.edu/ocean-photos/coral-gardener.

62. Austin Bowden-Kirby, "The Coral Gardener," *YouTube*, uploaded by Emma Robens, November 24, 2013. https://www.youtube.com/watch?v=8DXBQFzoMOs.

63. Joanna Macy and Chris Johnstone, *Active Hope: How to Face the Mess We're in without Going Crazy* (New World Library, 2012), 37.

64. Ibid., 3. See also Teresa Shewry, *Hope at Sea: Possible Ecologies in Oceanic Literature* (University of Minnesota Press, 2015), 5.

65. Les Kaufman (professor of biology, Boston University), interview by author, in person, Boston University, Boston, MA, April 5, 2017.

66. Austin Bowden-Kerby, *Best Practices Manual for Caribbean Acropora Restoration*, technical report, Puntacana Ecological Foundation, 2014. https://www.researchgate.net/publication/282245792_Best_Practices_Manual_for_Caribbean_Acropora_Restoration.

67. Tom Moore (coral restoration program manager, NOAA Restoration Center), interview by author, telephone, October 2, 2015.

68. Margaret W. Miller (research ecologist, NOAA/National Marine Fisheries Service, Southeast Fisheries Science Center), interview by author, telephone, September 16, 2015.

69. Miller, interview, September 16, 2015.

70. Michael Tlusty (director of research, New England Aquarium) and Andrew Rhyne (associate professor of marine biology, Roger Williams University), joint interview by author, in person, New England Aquarium, Boston, MA, May 11, 2016.

71. eBay, "Coral and Live Rock," accessed October 17, 2017. http://www.ebay.com.

72. Colin Foord (cofounder and codirector, Coral Morphologic), e-mail communication, November 14, 2017.

73. Smith, interview, July 18, 2016.

74. Ibid.

75. ADE Project, "ADE Mission Statement and Commitment," accessed July 12, 2017. http://www.adeproject.org/.

76. Smith, interview, July 18, 2016.

77. CITES, "CITES Quotas," accessed November 11, 2017. https://www.cites.org/eng/resources/quotas/2000/Indonesia.shtml.

78. Tlusty and Rhyne, interview, May 11, 2016. This was a joint interview, and the quotations weave together the interviewees' interjections.

79. Ibid. See also Andrew L. Rhyne et al., "Revealing the Appetite of the Marine Aquarium Fish Trade: The Volume and Biodiversity of Fish Imported into the United States," *PLoS ONE* 7 (2012): e35808; Andrew L. Rhyne et al., "Expanding Our Understanding of the Trade in Marine Aquarium Animals," *PeerJ* 5 (2017): e2949.

80. Colin Foord, e-mail communication, August 14, 2017.

81. See, e.g., calculations of "coral-colony-years" in chapter 4.

82. Ashley McCrea-Strub et al., "Understanding the Cost of Establishing Marine Protected Areas," *Marine Policy* 35 (2011): 1–9. See also Global Giving, "Corals for Conservation," accessed July 15, 2017. https://www.globalgiving.org/donate/26045/corals-for-conservation/info/. See also Katherine Ellison and Gretchen C. Daily, "Making Conservation Profitable," *Conservation* 4(2) (2003): 12–20.

83. See, e.g., Connor Cavanagh and Tor A. Benjaminsen, "Virtual Nature, Violent Accumulation: The 'Spectacular Failure' of Carbon Offsetting at a Ugandan National Park," *Geoforum* 56 (2014): 55–65; Noel Castree, "Commodifying What Nature?" *Progress in Human Geography* 27 (2003): 273–297.

84. Marxist geographers often distinguish between neoliberalism and neoliberalization: "It has become axiomatic among researchers that they are investigating a spatio-temporally variable process ('neoliberalisation') rather than a fixed and homogenous thing ('neoliberalism')." Noel Castree, "Neoliberalising Nature: The Logics of Deregulation and Reregulation," *Environment and Planning A* 40 (2008): 131–152, at 137. Castree nonetheless identifies the following characteristics as constituting "neoliberalism" when one abstracts from the multiple "neoliberalizations" in the world:

privatization, marketization, deregulation, reregulation (that is, the deployment of state policies to facilitate privatization and marketization of ever-wider spheres of social and environmental life), market proxies in the residual public sector (that is, the state-led attempt to run remaining public services along private-sector lines as "efficient" and "competitive" businesses), and the construction of flanking mechanisms in civil society (the state-led encouragement of civil society groups to provide services that interventionist states did, or could potentially, provide for (142). "Neoliberalism is simultaneously a social, environmental, and global project," Castree summarizes (143). See also Nik Heynen and Paul Robbins, "The Neoliberalization of Nature: Governance, Privatization, Enclosure and Valuation," *Capitalism Nature Socialism* 16 (2005): 1–4, at 2.

85. See, e.g., Noel Castree, "Neoliberalising Nature: Processes, Effects, and Evaluations," *Environment and Planning A* 40 (2008): 153–173; Jim Igoe and Dan Brockington, "Neoliberal Conservation: A Brief Introduction," *Conservation & Society* 5 (2007): 432–449; Rebecca Lave, "Neoliberalism and the Production of Environmental Knowledge," *Environment and Society: Advances in Research* 3 (2012): 19–38; Jessica Dempsey, *Enterprising Nature: Economics, Markets, and Finance in Global Biodiversity Politics* (Wiley-Blackwell, 2016); and James Fairhead, Melissa Leach, and Ian Scoones, "Green Grabbing: A New Appropriation of Nature?" *Journal of Peasant Studies* 39 (2012): 237–261.

86. See review by Castree, "Neoliberalising Nature: The Logics."

87. Rinkevich, e-mail communication, April 27, 2017.

88. Madeleine J. H. van Oppen et al., "Building Coral Reef Resilience through Assisted Evolution," *Proceedings of the National Academy of Sciences* 112 (2015): 2307–2313.

89. Van Oppen et al., "Shifting Paradigms."

90. Ibid., 4.

91. Miller, interview, September 16, 2015.

92. Rinkevich, "Climate Change."

93. Ibid., 123. See also Bart Linden and Baruch Rinkevich, "Elaborating an Eco-Engineering Approach for Stock Enhanced Sexually Derived Coral Colony," *Journal of Experimental Marine Biology and Ecology* 486 (2017): 314–321.

94. Rinkevich, "Climate Change," 119.

95. Rinkevich, interview, May 1, 2017. See also Elad Nehoray Rachmilovitz and Baruch Rinkevich, "Tiling the Reef: Exploring the First Step of an Ecological Engineering Tool That Promotes Phase-Shift Reversals in Coral Reefs," *Ecological Engineering* 105 (2017): 150–161.

96. Ken Anthony et al., "New Interventions Are Needed to Save Coral Reefs," *Nature Ecology & Evolution* 1 (2017): 1420–1422, at 1421.

97. Rinkevich, interview, May 1, 2017. See also Sweet, Ramsey, and Bulling, "Designer Reefs."

98. Elizabeth Cook, "Electric Reefs," *Alert Diver*, September/October 2007. http://www.globalcoral.org/_oldgcra/52_electricreefs_ads007.pdf.

99. BioRock International Corporation, "Biorock Technology." http://www.biorock.org/.

100. For example, it requires a power source to maintain its function, which makes for a less reliable operation in developing countries. See Rachel Fabian, Michael Beck, and Don Potts, "Reef Restoration for Coastal Defense: A Review," unpublished report, accessed November 25, 2017. http://www.car-spaw-rac.org/IMG/pdf/Reef_restoration_Coastal_Defense_report_Final-2.pdf. Additional concerns were raised with regard to recorded changes in the feeding behaviors of certain species, such as sharks, in response to the use of Biorock.Marcella P. Uchoa, Craig P. O'Connell, and Thomas J. Goreau, "The Effects of Biorock-Associated Electric Fields on the Caribbean Reef Shark (*Carcharhinus perezi*) and the Bull Shark (*Carcharhinus leucas*)," *Animal Biology* 67 (2017): 191–208.

101. Greg Rau, Elizabeth L. McLeod, and Ove Hoegh-Guldberg, "The Need for New Ocean Conservation Strategies in a High-Carbon Dioxide World," *Nature Climate Change* 2 (2012): 720–724.

102. See, e.g., John G. Shepherd, "Geoengineering the Climate: An Overview and Update," *Philosophical Transactions: Series A, Mathematical, Physical, and Engineering Sciences* 370 (2012): 4166–4175.

103. Mark Eakin (coordinator of Coral Reef Watch Program, NOAA), interview by author, telephone, November 16, 2015. See also Rau et al., "Need for New Ocean Conservation Strategies."

104. Ibid.

105. Florida Keys National Marine Sanctuary, "Corals Can Reproduce Sexually and Asexually," accessed November 8, 2017. https://floridakeys.noaa.gov/corals/reproduce.html. I will further discuss the corals' sexual reproduction and spawning in chapter 5.

106. Rinkevich, "Restoration Strategies," 245.

107. Dirk Petersen, Anne Wietheger, and Michaël Laterveer, "Influence of Different Food Sources on the Initial Development of Sexual Recruits of Reefbuilding Corals in Aquaculture," *Aquaculture* 277 (2008): 174–178. Rinkevich has also been studying and generating coral "chimeras": colonies that comprise tissues or cells of two or more genetically distinct individuals. The extent of chimerism in wild populations of reef corals is currently unknown. See Eneour Puill-Stephan et al., "Chimerism in Wild Adult Populations of the Broadcast Spawning Coral *Acropora millepora* on the Great Barrier Reef," *PLoS ONE* 4 (2009): e7751. See also chapter 5.

108. Austin Bowden-Kerby and Lisa Carne, "Thermal Tolerance as a Factor in Caribbean *Acropora* Restoration," *Proceedings of the 12th International Coral Reef Symposium*, Cairns, Australia, July 9–13, 2012.

109. Ibid.

110. Steve Palumbi (professor of biological sciences and director of Hopkins Marine Station, Stanford University), interview by author, in person, Waikiki, HI, June 24, 2016.

111. See chapter 5 for further discussion of the difference between this approach and assisted evolution as applied by Ruth Gates and Madeleine van Oppen.

112. Amelia Urry, "Coral Reefs Are in Trouble: Meet the People Trying to Rebuild Them," *Grist*, November 17, 2015. http://grist.org/science/coral-reefs-are-in-trouble-

meet-the-people-trying-to-rebuild-them/. See also Stephen R. Palumbi et al., "Mechanisms of Reef Coral Resistance to Future Climate Change," *Science* 344 (2014): 895–898.

113. "To test this, Palumbi's team transplanted corals from the cooler part of the reef to the warmest pool. They left them there for three years, and then subjected them to the same set of stress tests they had undergone three years earlier, raising the temperature to an extreme level for several days at a time. This time, the cool-adapted corals that had spent the last three years acclimating in warm pools did noticeably better than their cold-water clones. But the original warm-pool corals still outperformed their acclimated cousins, suggesting some older genetic advantage was also at work." Urry, "Coral Reefs Are in Trouble." See also Gergely Torda et al., "Rapid Adaptive Responses to Climate Change in Corals," *Nature Climate Change* 7 (2017): 627–636.

114. Urry, "Coral Reefs Are in Trouble."

115. Mary Hagedorn (director, MarineGEO Hawai'i, Smithsonian Conservation Biology Institute/Hawai'i Institute of Marine Biology), interview by author, telephone, September 24, 2015.

116. Florida Keys National Marine Sanctuary, "Corals Can Reproduce Sexually and Asexually."

117. Leonard Ho, "The Noble Work of Coral Reef Restoration," *Advanced Aquarist*, December 15, 2016. http://www.advancedaquarist.com/blog/noble-work-coral-reef-restoration.

118. SEZARC, accessed March 26, 2018. http://www.sezarc.org.

119. Linda Penfold (director, SEZARC), interview by author, Skype, July 26, 2016.

120. Foord, e-mail communication, August 17, 2017.

121. I first applied Michel Foucault's "great battle of pastorship" concept when discussing the struggles between zoo professionals and animal rights activists over who should speak for zoo animals. See Braverman, *Zooland*, 20–23.

122. Foord, e-mail communication, August 17, 2017.

123. Iliana B. Baums (associate professor of biology, The Pennsylvania State University; director, Center for Marine Science and Technology), interview by author, Skype, September 17, 2015.

124. Dirk Petersen (director of SExual COral Reproduction [SECORE]), interview by author, Skype, September 23, 2015.

125. Ibid.

126. SECORE, "Global Coral Reef Restoration Effort Launches in the Caribbean," April 11, 2017. http://www.secore.org/site/newsroom/article/global-coral-reef-restoration-effort-launches-in-the-caribbean.152.html. See also SECORE, "Global Coral Restoration Project," accessed July 5, 2017. http://www.secore.org/site/our-work/detail/global-coral-restoration-project.39.html.

127. Petersen, interview, September 23, 2015.

128. Braverman, *Wild Life*.

129. Frias-Torres, interview, June 23, 2016.

130. Moore, interview, October 2, 2015.

131. Ove Hoegh-Guldberg (director of Global Change Institute and professor of marine science, University of Queensland, Brisbane, Australia), interview by author, Skype, February 25, 2015. See also interchapter interview with Hoegh-Guldberg.

132. Rinkevich, interview, May 1, 2017.

133. Frias-Torres, interview, June 23, 2016. See also discussion of shifting baseline syndrome in chapter 2.

134. Ruth Gates (director, Hawai'i Institute of Marine Biology; president, International Society for Reef Studies), interview by author, in person, Coconut Island, Oahu, HI, July 1, 2016.

135. Hagedorn, interview, September 24, 2015.

136. Hagedorn, telephone communication, August 2016.

INTERVIEW. KEN NEDIMYER

1. "Live rock"—rock from the ocean, made from aragonite skeletons of dead corals—is used for saltwater aquariums because it acts as an ideal habitat for many forms of microscopic and macroscopic marine life who live on and inside the rock.

2. Nedimyer later updated me: "They do now, and they are fully on board," e-mail communication, October 13, 2017.

3. In our later communications in the aftermath of Hurricane Irma, Nedimyer told me that "we had that storm before any external gene bank was developed and fortunately the nursery survived and most of our genotypes are still there. . . . This was the ultimate test. My concern was that the sand would liquefy during the most intense parts of the storm, causing the sand anchors to let go and everything to float away. That didn't happen. The corals that had been properly prepared for the storm did extremely well, and the ones that had not been properly prepared were torn from the trees and were lost. [But] we only lost maybe five trees out of over five hundred, so that part of the technology proved itself. The outer reefs took a hard hit, with most soft corals, sponges, and algae getting ripped off the bottom and deposited in piles somewhere inshore. The elkhorn corals that we planted on the outer reefs did well, with about 90 percent survival rate, but the staghorn corals did not (no surprises there: the elkhorn is a lot more tolerant of getting banged around—they evolved to fill that niche, [while] the staghorn is an inshore coral). The inner reefs and the patch reefs seemed to fare better, but I have not been able to get out much to look around." Nedimyer, e-mail communication and telephone interview, October 13 and 17, 2017.

4. Nedimyer updated me that in the interim, they sent the larvae out to several other places, including Mote Marine Lab and the Georgia Aquarium, and that "this year all three organizations have successfully reared the larvae." Nedimyer, e-mail communication, October 13, 2017.

5. Nedimyer later updated: "Funny what a difference a year or two makes. . . . The projects are moving forward in other places around the globe, too. I'm realizing that the 'technology,' if properly applied, will work anywhere with any corals, so I don't

need to be a worldwide coral expert to be able to help people develop a program on the other side of the world." Ibid.

CHAPTER FOUR. CORAL LAW UNDER THREAT

1. R. Henry Weaver and Douglas A. Kysar, "Courting Disaster: Climate Change and the Adjudication of Catastrophe," *Notre Dame Law Review* 93 (2017): 295–356, at 296.

2. Linda Ross Meyer, "Catastrophe: Plowing Up the Ground of Reason," in *Law and Catastrophe*, edited by Austin Sarat, Lawrence Doglas, and Martha Umphrey (Stanford University Press, 2007), 21.

3. Lynn Margulis and Dorion Sagan, *What Is Life?* (University of California Press, 2000; originally published in 1995), 35.

4. Michel Foucault, *The History of Sexuality: An Introduction, vol. 1* (Vintage Books, 1990). On the application of biopolitics to nonhuman populations, see Cary Wolfe, *Before the Law: Humans and Other Animals in a Biopolitical Frame* (University of Chicago Press, 2012), as well as Irus Braverman (ed.), *Law, Animals, Biopolitics: Lively Legalities* (Routledge, 2016).

5. Indeed, the law is all over. Austin Sarat, "' . . . The Law Is All Over': Power, Resistance and the Legal Consciousness of the Welfare Poor," *Yale Journal of Law and the Humanities* 2 (1990): 343–379.

6. This definition, almost word-for-word, is from the United Nations, Division for Ocean Affairs and the Law of the Sea, "The United Nations Convention on the Law of the Sea (A Historical Perspective)," accessed November 4, 2017. http://www.un.org /depts/los/convention_agreements/convention_historical_perspective.

7. See interchapter interview with Roberts. However, in an unusual show of international restraint, in December 2017 a historic international treaty was signed between five nations with Arctic borders—Canada, Denmark (Greenland), Norway, Russia, and the United States—and other nations with industrialized fishing fleets—China, the European Union, Iceland, Japan, and South Korea—to ban commercial fishing in the high seas of the central Arctic Ocean for the next sixteen years at least. Until that point, no regulations had existed on fishing in this region because fishing was almost impossible there. But with the recent decrease in sea ice and the migration of fishing stocks toward the north, commercial fishing is becoming increasingly possible. This legally binding agreement covers 2.8 million square kilometers—an area slightly larger than the Mediterranean Sea—in the "high seas," which meant that, in theory, any country could have fished there. See Marine Ecosystem and Management (MEAM), "The Skimmer: A Very Quick Update on What's Going on in the Arctic Ocean Region and Why You Should Care," posted March 13, 2018. https://meam.openchannels.org /news/meam/skimmer-very-quick-update-whats-going-arctic-ocean-region-and-why-you-should-care.

8. Peter Sale (marine ecologist, professor emeritus, University of Windsor, Canada), written comments, August 7, 2017.

9. Callum Roberts, *The Ocean of Life* (Viking, 2012), 7–8.

10. Before my law professor colleagues—especially those from the critical legal studies tradition—roll their eyes to the sky and shout *"Gevald,"* I would like to clarify my use of the term *law*. By using this term, I do not mean to make sweeping generalizations about law as if it were homogeneous or uniform, and indeed I recognize that coral law is more like an overlapping mosaic or a patchwork of many different laws—formal and informal, binding and unbinding, hard and soft—that affect corals in myriad ways. Moreover, I realize that there are myriad laws that have no direct relationship with corals or even with the environment—for example, torts and property law and even flood insurance rules and policies—which nonetheless affect corals. Tax law, although not about the environment per se, probably impacts corals as much as environmental protection laws. Despite this recognition, the discussion in this chapter will be limited to those laws that affect corals more directly. I would like to thank Errol Meidinger for pushing me on this point.

11. But see Irus Braverman and Elizabeth R. Johnson (eds.), *Ocean Legalities: The Life and Law of the Sea* (Duke University Press, forthcoming).

12. "Increasingly, the realities of human experience emerge as the joint achievements of scientific, technical, and social enterprise: science and society, in a word, are co-produced, each underwriting the other's existence." Sheila Jasanoff, *States of Knowledge: The Co-production of Science and the Social Order* (Routledge, 2014), 17.

13. Mary Alice Coffroth (professor of geology, Graduate Program in Evolution, Ecology and Behavior, University at Buffalo, SUNY), interview by author, in person, and observation, Buffalo, NY, August 31, 2015.

14. Ibid.

15. According to the U.S. Geological Survey (USGS): "Extensive coral reefs are found in the waters of the United States and its territories. In the Atlantic Ocean, Gulf of Mexico, and the Caribbean Sea these include reefs off Florida, Texas, Puerto Rico, and the U.S. Virgin Islands. In the Pacific Ocean, they include those of the Hawaiian Islands, Wake Island, Johnston Atoll, the Northern Marianas, Saipan, Guam, Kingman Reef and Palmyra Atoll, Howland Island, Baker Island, Jarvis Island, and American Samoa. More than 60 percent of the Nation's coral reefs are found in the extended Hawaiian Island chain." USGS, "U.S. Coral Reefs—Imperiled National Treasures," accessed November 27, 2017. https://pubs.usgs.gov/fs/2002/fs025-02/.

16. American University Washington College of Law, "Coral Reefs," accessed July 7, 2017. https://www.wcl.american.edu/environment/iel/sup3.cfm.

17. NOAA, "What Is a Marine Protected Area?" revised July 6, 2017. https://oceanservice.noaa.gov/facts/mpa.html.

18. Roscoe Pound, "Law in Books and Law in Action," *American Law Review* 44(12) (1910): 12–36.

19. Magnuson-Stevens Fishery Conservation and Management Act (1976), 16 U.S.C. Sections 1801–1884. NOAA, Public Law 94–265.

20. NOAA, "Corals: Protecting Coral Reefs," revised July 6, 2017. https://oceanservice.noaa.gov/education/tutorial_corals/coral11_protecting.html.

21. Ibid.

22. Coral Reef Conservation Act of 2000, Pub. L. 106–562, Title II, December 23, 2000, 16 U.S.C. Section 6201 et seq.

23. NOAA, "Who We Are: Coral Reef Ecosystems: Valuable and Threatened," revised March 3, 2017. http://coralreef.noaa.gov/about/goalsandobjectives.html.

24. NOAA, "Recovery Plan Elkhorn Coral (*Acropora palmata*) and Staghorn Coral (*A. cervicornis*)," March 2015. https://data.nodc.noaa.gov/coris/library/NOAA /CRCP/project/2160/final_acropora_recovery_plan.pdf.

25. Gail Derr (ed.), "Researchers Explore Coral Resiliency in New Future Reef Laboratory," *AOML Keynotes* 21(1) (2017): 1–2. http://www.aoml.noaa.gov/keynotes /PDF-Files/Jan-Feb2017.pdf.

26. NOAA, "Coral Disease and Health Consortium," updated April 21, 2016. https://cdhc.noaa.gov.

27. NOAA, "Corals for the Future: A Research and Management Planning Agenda for Assistive Biological Enhancement to Reef Coral Sustainability" (report draft, cited with permission).

28. Peter Sale commented about my interview with Nedimyer: "Hidden is the fact that Nedimyer is obviously making a living through coral restoration, even if he laughs that he has never sold any corals. My own leftist tendencies tell me [that] it would be better if coral restoration did not become a business, but was in the hands of a reef management agency. But then NOAA has to be revamped so that it is working to sustain/restore reefs rather than applying regulations to people trying to do that!" Sale, written comments, August 7, 2017.

29. U.S. Coral Reef Task Force, *Handbook on Coral Reef Impacts: Avoidance, Mini-mization, Compensatory Mitigation, and Restoration*, December 2016. https://data .nodc.noaa.gov/coris/library/NOAA/CRCP/other/USCRTF/mitigation_handbook_ final_122216.pdf.

30. Ibid.

31. Noel Castree suggests deregulation and reregulation as two of six central tenets of neoliberalism. He defines *deregulation* as "the 'rollback' of state 'interference' in numerous areas of social and environmental life so that (i) state regulation is 'light touch' and (ii) more and more actors become self-governing within centrally prescribed frameworks and rules." His definition of *reregulation* is "the deployment of state policies to facilitate privatization and marketization of ever-wider spheres of social and environmental life." Noel Castree, "Neoliberalising Nature: The Logics of Deregulation and Reregulation," *Environment and Planning A* 40 (2008): 131–152, at 142.

32. Endangered Species Act (1973), 16 U.S.C. Sections 1531–1544.

33. Endangered and Threatened Species; Final Rule Listing Determinations for Elkhorn Coral and Staghorn Coral, 71 Fed. Reg. 26,852 (May 2, 2006) (codified at 50 D.F.R. §§ 223).

34. "Implementation of the latter restrictions have not been common and the case law is not extensive due to the challenges of establishing standing by third party liti-

gants." Kieran Suckling (director, Center for Biological Diversity), e-mail communication, December 2, 2017.

35. Sale, written comments, August 7, 2017.

36. National Marine Fisheries Service—NOAA, *Acropora* Recovery Team, "Recovery Plan for Elkhorn (*Acropora palmata*) and Staghorn (*A. cervicornis*) Corals," March 2015, 16. http://www.nmfs.noaa.gov/pr/recovery/plans/final_acropora_recovery_plan.pdf.

37. Ibid.

38. Sale, written comments, August 7, 2017.

39. U.S. Department of the Interior, "Trustees Settle National Resource Damage Claims Arising from February 2005 Cargo Ship Grounding Offshore Barbers Point, Oahu, Hawaii," updated February 14, 2017. https://www.doi.gov/restoration/news/mv-cape-flattery-grounding-settlement.

40. Clean Water Act (2008), 33 U.S.C. 1251, § 404.

41. Zac H. Forsman (researcher at Hawaiʻi Institute of Marine Biology and coral specialist for the State of Hawaiʻi), interview by author, telephone, March 1, 2016; and in person, Waikiki and Honolulu, HI, June 20 and 30, 2016.

42. Aurora Fredriksen, "Valuing Species: The Continuities between Non-market and Market Valuations in Biodiversity Conservation," *Valuation Studies* 5 (2017): 39–59, at 41. See also Thom van Dooren, "Authentic Crows: Identity, Captivity and Emergent Forms of Life," *Theory, Culture & Society* 33(7–8) (2016): 29–52.

43. Fredriksen, "Valuing Species," 51.

44. Sian Sullivan, "On 'Natural Capital,' 'Fairy-Tales' and Ideology," *Development and Change* 48 (2017): 397–423, at 398.

45. Endangered and Threatened Wildlife and Plants: Final Listing Determinations on Proposal to List 66 Reef-Building Coral Species and to Reclassify Elkhorn and Staghorn Corals, 79 Fed. Reg., Issue 195 (September 10, 2014).

46. NOAA Fisheries, "Corals," August 25, 2014. http://www.nmfs.noaa.gov/pr/species/invertebrates/corals.htm.

47. Miyoko Satashita (oceans director and senior counsel, Center for Biological Diversity), interview by author, telephone, October 8, 2015.

48. These two sentences are near-verbatim paraphrases of Jonathan Lovvorn, "Climate Change beyond Environmentalism Part I: Intersectional Threats and the Case for Collective Action," *Georgetown Environmental Law Review* 29 (2016): 1–67, at 13.

49. Alison Moulding (natural resource specialist, Fisheries Southeast Regional Office Protected Resources Division, NOAA), interview by author, telephone, December 16, 2015.

50. Margaret W. Miller (research ecologist, NOAA/National Marine Fisheries Service, Southeast Fisheries Science Center), interview by author, telephone, September 16, 2015.

51. Ibid.

52. Nilda Jiménez-Marrero (biologist, Puerto Rico Department of Natural Resources), interview by author, in person, Boquerón, PR, January 20, 2014; and by telephone, October 7, 2014.

53. Andrew Rhyne (assistant professor of marine biology), interview by author, in person, New England Aquarium, Boston, MA, May 11, 2016 (joint interview with Michael Tlusty).

54. Jennifer Moore (marine conservation program manager, NOAA), interview by author, telephone, September 30, 2015.

55. Ibid.

56. Rhyne, interview, May 11, 2016.

57. For the role of classification in science, see, e.g., Geoffrey Bowker and Susan L. Star, *Sorting Things Out: Classification and Its Consequences* (MIT Press, 1999).

58. Moore, interview, September 30, 2015.

59. Geographer Aurora Fredriksen discusses Scottish wildcat conservation, which, she argues, revolves around discerning "whether an individual wild-living cat in Scotland is a 'pure' Scottish wildcat or a hybrid." Aurora Fredriksen, "Of Wildcats and Wild Cats: Troubling Species-Based Conservation in the Anthropocene," *Environment and Planning D: Society and Space* 34 (2016): 689–705. As Fredriksen explains in a later article: "This is not simply a matter of blind ideology (though there is some of that in the mix), but also a pragmatic response to the current legal environment which affords strict protection for Scottish wildcats but allows hybrid and feral cats to be shot on sight. Indeed, such constraints in the wider governance of biodiversity conservation, prominently including national and international legal regimes, generally afford protection for animals only at the level of clearly defined species, thus acting as another site where the value of animals and other organisms is performatively located at the level of species." Fredriksen, "Valuing Species," 40. See also van Dooren, "Authentic Crows"; Irus Braverman, *Wild Life: The Institution of Nature* (Stanford University Press, 2015).

60. NOAA, "Recovery Plan for Acroporas."

61. National Marine Fisheries Service—NOAA, Acropora Biological Review Team, "Atlantic Acropora Status Review Document. Report to National Marine Fisheries Service, Southeast Regional Office," March 3, 2005, 19. http://www.nmfs.noaa.gov/pr/pdfs/statusreviews/corals.pdf.

62. Miller, interview, September 16, 2015.

63. Ibid.

64. Forsman, interview, June 30, 2016.

65. Ann F. Budd and Hector M. Guzman, "*Siderastrea glynni*, a New Species of Scleractinian Coral (Cnidaria, Anthozoa) from the Eastern Pacific," *Proceedings of the Biological Society of Washington* 107 (1994): 591–599.

66. Douglas Fenner, "Mass Bleaching Threatens Two Coral Species with Extinction," *Reef Encounters* 29 (2001): 9–10.

67. Endangered and Threatened Wildlife and Plants: Final Rule to List the Dusky Sea Snake and Three Foreign Corals under the Endangered Species Act, 80 FR 60560 (codified at 50 CFR §§ 223 and 224), October 7, 2015.

68. For a detailed discussion of the Red List, see Irus Braverman, "En-Listing Life: Red Is the Color of Threatened Species Lists," in *Critical Animal Geographies*, edited by Rosemary-Claire Collard and Kathryn Gillespie (Routledge/Earthscan, 2015). See also Irus Braverman, "The Regulatory Life of Threatened Species Lists," in *Lively Legalities: Animals, Biopolitics, Law*, edited by Irus Braverman (Routledge, 2016), 18–36.

69. Shaye Wolf explained that the rationale behind the massive petition was that these coral species were already identified as imperiled in 2008 by the IUCN Red List. Shaye Wolf (climate change director, Center for Biological Diversity), interview by author, Skype, November 4, 2015.

70. Zac H. Forsman et al., "An ITS Region Phylogeny of *Siderastrea* (Cnidaria: Anthozoa): Is *S. glynni* Endangered or Introduced?" *Coral Reefs* 24 (2005): 343–347.

71. In conservation terms, "introduced" species are defined as "non-native" species that occur as a result of human activity. Introduced species can become "invasive," but not all invasive species were introduced by humans. These specific terms have generated considerable debates among conservation scientists.

72. Forsman, interview, June 30, 2016. See also R. E. Brainard et al., "Status Review Report of 82 Candidate Coral Species Petitioned under the U.S. Endangered Species Act," *Department of Commerce NOAA Technical Memo* (NOAA-TM-NMFS-PIFSC-27, 2011), 530. https://www.pifsc.noaa.gov/library/pubs/tech/NOAA_Tech_Memo_PIFSC_27.pdf.

73. Todd C. LaJeunesse, Zac H. Forsman, and Drew C. Wham, "An Indo-West Pacific 'Zooxanthella' Invasive to the Western Atlantic Finds Its Way to the Eastern Pacific Via an Introduced Caribbean Coral," *Coral Reefs* 35 (2016): 577–582.

74. See also Peter W. Glynn et al., "The True Identity of *Siderastrea glynni* Budd & Guzmán, 1994, a Highly Endangered Eastern Pacific Scleractinian Coral," *Coral Reefs* 35 (2016): 1399–1404.

75. The IUCN Red List of Threatened Species, *Siderastrea glynni*, 2008. http://www.iucnredlist.org/details/133121/0.

76. See, e.g., Andrew T. Guzman and Timothy L. Meyer, "International Soft Law," *Journal of Legal Analysis* 2 (2010): 171–225; Gregory Shaffer and Mark A. Pollack, "Hard vs. Soft Law: Alternatives, Complements and Antagonists in International Governance," *Minnesota Law Review* 94 (2010): 706–799.

77. This story brings to mind a long-standing academic debate in science and technology studies about whether the bureaucracy-heavy law always and inevitably lags behind the tech-savvy science. See, e.g., Sheila Jasanoff, *Science at the Bar: Law, Science, and Technology in America* (Harvard University Press, 1995), 11–12.

78. Michel Foucault, *The History of Sexuality* (Vintage Books, 1990), 143. Geographer Elizabeth Hennessy's work on giant tortoise conservation demonstrates the slipperiness of biological processes. In her words, "In contradistinction to celebrations of the seemingly limitless possibilities of modern biology to remake genealogy, attention to the slow, provisional processes of knowledge production in this case stresses how difficult it is for people to achieve control over biological processes."

Elizabeth Hennessy, "Producing 'Prehistoric' Life: Conservation Breeding and the Remaking of Wildlife Genealogies," *Geoforum* 49 (2013): 71–80, at 72.

79. Intergovernmental Panel on Climate Change (IPCC), "Climate Change 2014 Synthesis Report Summary for Policymakers," accessed November 25, 2017. https://www.ipcc.ch/pdf/assessment-report/ar5/syr/AR5_SYR_FINAL_SPM.pdf. This is the fifth IPCC report. For a related discussion and critique of the fourth IPCC report from 2007, see Jessica O'Reilly, "Glacial Dramas: Typos, Projections, and Peer Review in the Fourth Assessment of the Intergovernmental Panel on Climate Change," in *Climate Cultures: Anthropological Perspectives on Climate Change*, edited by Jessica Barnes and Michael R. Dove (Yale University Press, 2015), 107–126.

80. Moore, interview, September 30, 2015.

81. Part and parcel of mainstream scientific discourse, doubt and uncertainty have more typically played out to the disadvantage of environmental policy. See Naomi Oreskes and Erik M. Conway, *Merchants of Doubt: How a Handful of Scientists Obscured the Truth on Issues from Tobacco Smoke to Global Warming* (Bloomsbury Press, 2010). See also Joshua P. Howe, *Behind the Curve: Science and the Politics of Global Warming* (University of Washington Press, 2014).

82. Braverman, *Wild Life*. See also Rafi Youatt, "Counting Species: Biopower and the Global Biodiversity Census," *Environmental Values* 17 (2008): 393–417.

83. On the significance of numbers in listing processes, see Braverman, "En-Listing Life" and "Regulatory Life of Threatened Species Lists."

84. Chapter 5 discusses this in more depth.

85. Miller, interview, September 16, 2015.

86. Moulding, interview, December 16, 2015.

87. See chapter 5. See also Scott Gilbert, Jan Sapp, and Alfred I. Tauber, "A Symbiotic View of Life: We Have Never Been Individuals," *The Quarterly Review of Biology* 87 (2012): 325–341.

88. Moulding, interview, December 16, 2015.

89. Ibid.

90. Ruth Gates (director, Hawai'i Institute of Marine Biology; president, International Society for Reef Studies), interview by author, Skype, January 25, 2016.

91. Moulding, interview, December 16, 2015.

92. Miller, interview, September 16, 2015. The use of inference in lieu of accurate quantitative data, although seemingly unscientific, is in fact quite common in the work of conservation scientists. During my research on how endangerment is calculated, I came across sophisticated models and algorithms developed by scientists to calculate what they refer to as "fuzzy numbers" and "unknown knowns." For a detailed discussion of fuzzy numbers in the context of threatened species lists, see Irus Braverman, "Anticipating Endangerment: The Biopolitics of Threatened Species Lists," *BioSocieties* 12 (2017): 132–157.

93. Moore, interview, September 30, 2015.

94. Ibid.

95. Ibid.

96. Richard J. Lazarus, "Super Wicked Problems and Climate Change: Restraining the Present to Liberate the Future," *Cornell Law Review* 94 (2009): 1153–1233, at 1159. See also Frank P. Incropera, *Climate Change: A Wicked Problem—Complexity and Uncertainty at the Intersection of Science, Economics, Politics, and Human Behavior* (Cambridge University Press, 2015).

97. Lovvorn, "Climate Change beyond Environmentalism," 11.

98. American University Washington College of Law, "Coral Reefs."

99. International Coral Reef Initiative, "Global Coral Reef Monitoring Network (GCRMN)," updated January 16, 2017. http://www.icriforum.org/gcrmn.

100. International Coral Reef Initiative, "International Year of the Reef (IYOR)," accessed July 7, 2017. http://www.icriforum.org/about-icri/iyor.

101. The World Heritage Convention. https://whc.unesco.org/en/convention/.

102. Dennis Normile, "Almost All of the 29 Coral Reefs on U.N. World Heritage List Damaged by Bleaching," *Science*, June 26, 2017. http://www.sciencemag.org/news/2017/06/almost-all-29-coral-reefs-un-world-heritage-list-damaged-bleaching.

103. Scott F. Heron, Mark Eakin, and Fanny Douvere, "Impacts of Climate Change on World Heritage Coral Reefs: A First Global Scientific Assessment," UNESCO World Heritage Centre, June 23, 2017. http://whc.unesco.org/en/news/1676.

104. Normile, "Almost All."

105. Convention on International Trade in Endangered Species of Wild Fauna and Flora (CITES), March 3, 1973, 27 U.S.T. 1087; T.I.A.S. no. 8249; 993 U.N.T.S. 243.

106. I am grateful to Doug Kysar for this qualification.

107. J Murray Roberts (professor of marine biology, University of Edinburgh), interview by author, in person, Edinburgh, UK, April 20, 2016.

108. Roberts, interview, Skype, January 28, 2016.

109. *Foster v. Washington Department of Ecology* (no. 14–2–25295–1 SEA), Petitioners' Opening Brief, 17. https://static1.squarespace.com/static/571d109b04426270152febe0/t/576080e81bbee08251f2841c/1465942249247/ATL.Opening+Brief.Final_.3.16.14.pdf.

110. 16 Wash. Reg. 19–047 (Sept. 15, 2016) (codified in Wash. Admin. Code §§ 173–441, 173–442). http://lawfilesext.leg.wa.gov/law/wsr/2016/19/16–19–047.htm. See also Weaver and Kysar, "Courting Disaster," 345.

111. First Amended Complaint, *Juliana v. United States*, no. 6:15-cv-01517 (D. Or. Sept. 10, 2015).

112. Weaver and Kysar, "Courting Disaster," 346–347.

113. First Amended Complaint, *Juliana v. United States*, no. 6:15-cv-01517 (D. Or. Sept. 10, 2015), 40.

114. For an explanation of parts per million, see Scripps Institution of Oceanography, "The Keeling Curve: What Does this Number Mean?" posted May 12, 2015. https://scripps.ucsd.edu/programs/keelingcurve/2015/05/12/what-does-this-number-mean/.

115. First Amended Complaint, *Juliana v. United States*, 30–31, at 86.

116. *Juliana v. United States*, 217 F. Supp. 3d 1224, 1250 (D. Or. 2016).

117. Julia Olson (executive director and chief legal counsel, Our Children's Trust), interview by author, telephone, November 6, 2017.

118. Ibid. Olson's ideas of law seem to draw on conceptions of "unwritten fundamental law" developed in political thought around the time of the American Revolution. Thomas C. Grey, "Origins of the Unwritten Constitution: Fundamental Law in American Revolutionary Thought," *Stanford Law Review* 30 (1978): 843–893.

119. For updates on this case, see Our Children's Trust, https://www.ourchildrenstrust.org/us/federal-lawsuit/.

120. Australian Government—Great Barrier Reef Marine Park Authority (GBRMPA), "Crown-of-Thorns Starfish," accessed June 13, 2017. http://www.gbrmpa.gov.au/about-the-reef/animals/crown-of-thorns-starfish.

121. Australian Government, "Crown-of-Thorns Starfish." Historian Jan Sapp documented the debate that ensued over the primary reason for the COTS outbreak, asking whether such outbreaks were part of the natural cycle or the result of human activities. Jan Sapp, *What Is Natural? Coral Reef Crisis* (Oxford University Press, 1999). See also James Bowen, *The Coral Reef: From Discovery to Decline* (Springer, 2015), 141–143.

122. Living Oceans Foundation, "Crown of Thorns Starfish (COTS)," accessed June 20, 2017. https://www.livingoceansfoundation.org/science/crown-of-thorns-starfish/.

123. Australian Government—Great Barrier Reef Marine Park Authority, *Crown-of-Thorns Starfish Control Guidelines*, 2014. http://www.gbrmpa.gov.au/__data/assets/pdf_file/0006/185298/COTS-control-guidelines.pdf.

124. David Wachenfeld (director of reef recovery, GBRMPA), interview by author, Skype, January 4, 2016.

125. Irus Braverman, "The Life and Law of Corals: Breathing Meditations," in *Handbook of Research Methods in Environmental Law*, edited by Victoria Brooks and Andreas Philippopoulous-Mihalopoulos (Edward Elgar, 2017), 458–481.

126. *The Guardian*, "Robotic Killer Being Trialed to Rid Great Barrier Reef of Crown-of-Thorns Starfish," September 3, 2015. https://www.theguardian.com/environment/2015/sep/03/robotic-killer-being-trialled-to-rid-great-barrier-reef-of-crown-of-thorns-starfish.

127. Matthew Dunbabin (principal research fellow, Department of Science and Engineering, Queensland University of Technology, Australia), interview by author, Skype, January 20, 2016.

128. Ibid.

129. Ibid.

130. Ibid.

131. See also Irus Braverman, "Robotic Life in the Deep Sea: Deploying Killer (and Other) Robots to Make Live," in Braverman and Johnson, *Ocean Legalities*.

132. Olson, interview, November 6, 2017.

133. Ibid.

INTERVIEW. J MURRAY ROBERTS

1. See, e.g., Murray Roberts et al., "Cold-Water Corals in an Era of Rapid Global Change: Are These the Deep Ocean's Most Vulnerable Ecosystems?" in *The Cnidaria: Past, Present and Future* (Springer International, 2016), 593–606.

2. Similarly, in an interview we held during the second UNCLOS Preparatory Committee meeting on the conservation and sustainable use of marine biological diversity beyond areas of national jurisdiction, Alistair Graham stressed: "When [in the 1990s] people found that there were perfectly intact and coherent coral communities in what was generally regarded as impossibly deep water they just sort of went, 'Holy shit, what's this?' That was quite something. From that point of view, the coral has been really important for the whole deep-sea conservation game, because that was the entry point for quite a lot of people and [for] quite a lot of the scientific community. [This meant that] we now have to completely rewrite the book. Basically, it meant you have to manage deep-sea systems as biological systems rather than just deserts of minerals, water, [and] chemicals." Alistair Graham (high seas policy advisor, World Wildlife Fund International; NGO nominee on the Australian government delegation, UNCLOS PrepCom), interview by author, in person, United Nations Headquarters, New York, NY, August 30, 2016.

CHAPTER FIVE. THE CORAL HOLOBIONT

1. Anna L. Tsing, *The Mushroom at the End of the World: On the Possibility of Life in Capitalist Ruins* (Princeton University Press, 2015), 28.

2. Eugene Rosenberg and Ilana Zilber-Rosenberg, *The Hologenome Concept: Human, Animal and Plant Microbiota* (Springer, 2010), 4.

3. Ibid., 4. Another study conducted in 1999 found that in a stable state, the density count of *Symbiodinium* is more than two million per square centimeter. Cited in James Bowen, *The Coral Reef Era: From Discovery to Decline* (Springer, 2015), 156. For this reason, Haraway argues that the term *host*, as well as the host-symbiont dichotomy, are odd, suggesting instead that "all the partners making up holobionts are symbionts to each other." Donna J. Haraway, *Staying with the Trouble: Making Kin in the Chthulucene* (Duke University Press, 2016), 67.

4. Rosenberg and Zilber-Rosenberg, *Hologenome Concept*, 5.

5. Paola Furla, Sophie Richier, and Denis Allemand, "Physiological Adaptation to Symbiosis in Cnidarians," in *Coral Reefs: An Ecosystem in Transition*, edited by Zvy Dubinsky and Noga Stambler (Springer, 2014), 187–195, at 187.

6. Ibid., 187.

7. Ruth Gates (director, Hawai'i Institute of Marine Biology; president, International Society for Reef Studies), interview by author, Skype, January 25, 2016.

8. For a broader discussion of the "molecular turn" in conservation, see Elizabeth Hennessy, "The Molecular Turn in Conservation: Genetics, Pristine Nature, and the Rediscovery of an Extinct Species of Galapagos Giant Tortoise," *Annals of the Association of American Geographers* 105 (2015): 87–104.

9. Rosenberg and Zilber-Rosenberg, *Hologenome Concept*, 169. See also Scott F. Gilbert, Jan Sapp, and Alfred I. Tauber, "A Symbiotic View of Life: We Have Never Been Individuals," *The Quarterly Review of Biology* 87 (2012): 325–341, at 334; Maurizio Meloni, "How Biology Became Social, and What It Means for Social Theory," *The Sociological Review* 62 (2014): 593–614.

10. Lynn Margulis, "Symbiogenesis and Symbioticism," in *Symbiosis as a Source of Evolutionary Innovation: Speciation and Morphogenesis*, edited by Lynn Margulis and René Fester (MIT Press, 1991), 1–14.

11. Lynn Margulis and Dorion Sagan, *What Is Life?* (University of California Press, 2000; originally published in 1995), 236.

12. Francisco Carrapiço, "Can We Understand Evolution without Symbiogenesis?" in *Reticulate Evolution: Symbiogenesis, Lateral Gene Transfer, Hybridization and Infectious Heredity*, edited by Nathalie Gontier (Springer, 2015), 81–105.

13. Gilbert et al., "Symbiotic View of Life," 326.

14. Ibid.

15. Rosenberg and Zilber-Rosenberg, *Hologenome Concept*, 82.

16. This is the terminology used by Eldredge and Gould. Stephen Jay Gould and Niles Eldredge, "Punctuated Equilibria: An Alternative to Phyletic Gradualism," in *Models in Paleobiology*, edited by Thomas J. M. Schopf (Freeman, Cooper, 1972), 82–115. See also Ilana Zilber-Rosenberg and Eugene Rosenberg, "Role of Microorganisms in the Evolution of Animals and Plants: The Hologenome Theory of Evolution," *FEMS Microbiology Reviews* 32 (2008): 723–735; Stuart A. Newman, "Developmental Plasticity and Organismal Ingenuity Challenge Darwin's Theory," *Huffington Post*, April 22, 2015. http://www.huffingtonpost.com/stuart-a-newman/developmental-plasticity-and-organismal-ingenuity-challenge-darwins-theory_b_6701678.html.

17. Iliana B. Baums (associate professor of biology, The Pennsylvania State University; director, Center for Marine Science and Technology), interview by author, Skype, September 17, 2015.

18. Ibid.

19. Ibid. See also NOAA, "Coral Reef Information Glossary: Genet," accessed December 2, 2017. https://definedterm.com/a/definition/184759.

20. Mikhail Matz (professor of integrative biology, University of Texas at Austin), interview by author, Skype, May 2, 2016.

21. John Finnerty (associate professor of biology, Boston University), interview by author, in person, Boston University, Boston, MA, April 5, 2017.

22. Karl Marx, *Capital*, vol. 1, chapter 14, section 5, fn. 41 (Online Library of Liberty, 1909; originally published in 1867).

23. Ibid., chapter 13, at 366.

24. Leah Reshef et al., "The Coral Probiotic Hypothesis," *Environmental Microbiology* 8 (2006): 2068–2073. See also Rosenberg and Zilber-Rosenberg, *Hologenome Concept*, vii.

25. Rosenberg and Zilber-Rosenberg, *Hologenome Concept*, 1.

26. Nancy Knowlton and Forest Rohwer, "Multispecies Microbial Mutualisms on Coral Reefs: The Host as a Habitat," *The American Naturalist* 162 (Supplement 4) (2003): S51–S62.

27. David Sloan Wilson and Elliott Sober, "Reviving the Superorganism," *Journal of Theoretical Biology* 136 (1989): 337–356, at 337. See also Matt Haber, "Colonies Are Individuals: Revisiting the Superorganism Revival," in *From Groups to Individuals: Evolution and Emerging Individuality*, edited by Frédéric Bouchard and Philippe Hunema (MIT Press, 2013), 195–217; Thomas C. G. Bosch and Margaret J. McFall-Ngai, "Metaorganisms as the New Frontier," *Zoology* 114 (2011): 185–190.

28. See, e.g., Bert Hölldobler and Edward O. Wilson, *The Ants* (Harvard University Press, 1990).

29. James E. Lovelock, "Hands Up for the Gaia Hypothesis," *Nature* 344 (1990): 100–102.

30. Thomas C. G. Bosch, and David J. Miller, *The Holobiont Imperative: Perspectives from Early Emerging Animals* (Springer, 2016).

31. Margaret W. Miller (research ecologist, NOAA/National Marine Fisheries Service, Southeast Fisheries Science Center), interview by author, telephone, September 16, 2015.

32. Anonymous interview by author, May 2016.

33. See, e.g., the discussion of *Siderastrea glynni* in chapter 4.

34. Michel Pichon (tropical marine consultant and honorary research associate, Museum of Tropical Queensland), interview by author, in person, Waikiki, HI, June 20, 2016.

35. John E. N. "Charlie" Veron, *Corals of the World* (Australian Institute of Marine Science, 2000); Corals of the World, "Welcome to Corals of the World," accessed June 20, 2017. http://www.coralsoftheworld.org.

36. Zac H. Forsman (researcher, Hawai'i Institute of Marine Biology; coral specialist, State of Hawai'i), interview by author, in person, Waikiki, HI, June 20, 2016.

37. This quote and the next two quotes are from John "Charlie" Veron (former chief scientist, Australian Institute of Marine Science), interview by author, in person, Waikiki, HI, June 22, 2016.

38. See chapter 4.

39. Nicole Fogarty (assistant professor, Nova Southeastern University Oceanographic Center), interview by author, Skype, June 15, 2017.

40. Steven V. Vollmer and Stephen R. Palumbi, "Hybridization and the Evolution of Reef Coral Diversity," *Science* 296 (2002): 2023–2025.

41. Ibid.

42. Madeleine J. H. van Oppen et al., "Examination of Species Boundaries in the *Acropora cervicornis* Group (Scleractinia, Cnidaria) Using Nuclear DNA Sequence Analyses," *Molecular Ecology* 9 (2000): 1363–1373.

43. David J. Miller and Madeleine J. H. van Oppen, "A 'Fair Go' For Coral Hybridization," *Molecular Ecology* 12 (2003): 805–807, at 807.

44. Ibid., 806.

45. See discussion about the dusky seaside sparrow in Irus Braverman, *Wild Life: The Institution of Nature* (Stanford University Press, 2015), 165–166.

46. NOAA, "Atlantic Acropora Status Review," March 3, 2005. http://www.nmfs .noaa.gov/pr/pdfs/statusreviews/corals.pdf. See also chapter 4.

47. See interchapter interview with Nedimyer.

48. Bette L. Willis et al., "The Role of Hybridization in the Evolution of Reef Corals," *Annual Review of Ecology, Evolution, and Systematics* 37 (2006): 489–517.

49. Madeleine J. H. van Oppen (professor, University of Melbourne; senior principal research scientist, Australian Institute of Marine Science), interview by author, Skype, October 15, 2015. See also discussion of purity in Scottish wildcats, which revolves around discerning "whether an individual wild-living cat in Scotland is a 'pure' Scottish wildcat or a hybrid." Aurora Fredriksen, "Of Wildcats and Wild Cats: Troubling Species-Based Conservation in the Anthropocene," *Environment and Planning D: Society and Space* 34 (2016): 689–705. See also Thom van Dooren, "Authentic Crows: Identity, Captivity and Emergent Forms of Life," *Theory, Culture & Society* 33(2) (2016): 29–52.

50. Stuart Newman (professor of cell biology and anatomy, New York Medical College), e-mail communication, May 22, 2017.

51. Jente Ottenburghs et al., "Hybridization in Geese: A Review," *Frontiers in Zoology* 13 (2016): 1–9.

52. Dietmar Zinner, Michael L. Arnold, and Christian Roos, "The Strange Blood: Natural Hybridization in Primates," *Evolutionary Anthropology: Issues, News, and Reviews* 20 (2011): 96–103.

53. Carlos López-Larrea (ed.), *Self and Nonself* (Springer, 2012), 53.

54. Douglas Zook, "Symbiosis—Evolution's Co-Author," in *Reticulate Evolution: Symbiogenesis, Lateral Gene Transfer, Hybridization and Infectious Heredity*, edited by Nathalie Gontier (Springer, 2015), 41–80. Rinkevich, too, is passionate about chimeras and believes that they are considerably more prominent and important than scientists think. Baruch Rinkevich (senior scientist, National Institute of Oceanography, Haifa, Israel), interview by author, telephone, May 1, 2017.

55. López-Larrea, *Self and Nonself*, 54.

56. Gates, interview, January 25, 2016.

57. James R. Lupski, "Genome Mosaicism—One Human, Multiple Genomes," *Science* 341 (2013): 358–359.

58. T. A. Jones and Thomas A. Monaco, "A Role for Assisted Evolution in Designing Native Plant Materials for Domesticated Landscapes," *Frontiers in Ecology and the Environment* 7 (2009): 541–547.

59. Madeleine J. H. van Oppen et al., "Shifting Paradigms in Restoration of the World's Coral Reefs," *Global Change Biology* 3 (2017): 3437–3448.

60. Van Oppen, interview, October 15, 2015.

61. Gates, interview, January 25, 2016.

62. Van Oppen et al., "Shifting Paradigms," 6.

63. Ibid.

64. Van Oppen, interview, October 15, 2015.

65. Gates, interview, January 25, 2016. Relatedly, the first documented winners-losers discussion in the coral context was in 2001 by Yossi Loya and others. Yossi Loya et al., "Coral Bleaching: The Winners and the Losers," *Ecology Letters* 4 (2001): 122–131, at 126. See also Robert van Woesik, Akiyuki Irikawa, and Yossi Loya, "Coral Bleaching: Signs of Change in Southern Japan," in *Coral Health and Disease*, edited by Eugene Rosenberg and Yossi Loya (Springer, 2013), 119–141.

66. Gates, interview, January 25, 2016.

67. Gates, e-mail communication, August 16, 2017. But see the definition of *super coral* in Colin Foord, "A Hybrid Future: The Corals of Miami," YouTube, uploaded by TEDxMia, October 5, 2011. https://www.youtube.com/watch?v=3X3uCnEN7lE.

68. Van Oppen, interview, October 15, 2015.

69. Van Oppen et al., "Shifting Paradigms," 8.

70. Madeleine J. H. van Oppen et al., "Building Coral Reef Resilience through Assisted Evolution," *Proceedings of the National Academy of Sciences* 112 (2015): 2307–2313. See also Rachel A. Levin et al., "Engineering Strategies to Decode and Enhance the Genomes of Coral Symbionts," *Frontiers in Microbiology* 8 (2017): 1–11.

71. As I was preparing the book for final proofs, Ken Anthony of the Australian Institute of Marine Science, along with seventeen colleagues, published an article that calls on coral scientists to consider precisely such strategies, and others, "against the alternative of continual decline and increasingly insufficient and costly management." In their words, "We believe that assisted evolution and synthetic biology could offer more opportunities than risks for climate-hardening coral reefs, as long as such technologies are developed and deployed under a stringent and adaptive framework that includes extensive societal consultation." Ken Anthony et al., "New Interventions Are Needed to Save Coral Reefs," *Nature Ecology & Evolution* 1 (2017): 1420–1422.

72. Mikhail Matz (professor of integrative biology, University of Texas at Austin), interview by author, Skype, May 2, 2016.

73. Baruch Rinkevich remarks along the same lines: "This is not 'assisted' evolution. This is full manipulation of evolution—[which is similar to] the de-extinction approach." Rinkevich, e-mail communication, May 10, 2017.

74. Matz, interview, May 2, 2016.

75. Ibid.

76. Anthony et al., "New Interventions Are Needed," 1421 (citations omitted).

77. Groves B. Dixon et al., "Coral Reefs: Genomic Determinants of Coral Heat Tolerance across Latitudes," *Science* 348 (2015): 1460–1462 (abstract).

78. Sandy Tudhope (professor, School of GeoSciences, University of Edinburgh), interview by author, Skype, January 29, 2016.

79. Ibid.

80. Ibid.

81. "The sexuality of corals—whether they are hermaphrodite or separately sexed—tends to be generally consistent within species and genera, although there are exceptions.

There is sometimes also geographic variation within species." See Australian Institute of Marine Science, "Coral Fact Sheets," accessed March 26, 2018. https://coral.aims .gov.au/info/reproduction-sexual.jsp. See also SECORE, "Coral Reproduction," accessed March 26, 2018. http://www.secore.org/site/corals/detail/coral-reproduction .15.html.

82. Don R. Levitan et al., "Genetic, Spatial, and Temporal Components of Precise Spawning Synchrony in Reef Building Corals of the *Montastraea annularis* Species Complex," *Evolution* 65 (2011): 1254–1270. In a related discussion about synchrony in marine microorganisms, Astrid Schrader asks: "Did you know that microbes get jet-lagged?" Astrid Schrader, "Marine Microbiopolitics: Haunted Microbes before the Law," in *Ocean Legalities: The Law and Life of the Sea*, edited by Irus Braverman and Elizabeth R. Johnson (Duke University Press, forthcoming).

83. Levitan, "Genetic, Spatial, and Temporal Components," 1261.

84. Nicole Fogarty, telephone interview, May 1, 2017.

85. Levitan, "Genetic, Spatial, and Temporal Components," 1266.

86. Ibid.

87. "Coral Spawning Caught on Tape in Florida Keys (Video)," *The Huffington Post*, August 30, 2013. http://www.huffingtonpost.com/2013/08/29/coral-spawning_n_3839188.html.

88. Linda Penfold (director, South-East Zoo Alliance for Reproduction and Conservation [SEZARC]), interview by author, Skype, July 26, 2016.

89. Ibid.

90. Ibid.

91. Harvard-Smithsonian Center for Astrophysics, "Modern Techniques to Conserve Coral Reefs," accessed November 27, 2017. https://www.cfa.harvard.edu/hea /sws/docs/HagedornCryoProgram.pdf.

92. Ibid.

93. Ibid.

94. For a further discussion of gene banks, or "frozen zoos," and their critique by many in the conservation community, see Braverman, *Wild Life*, 99–101.

95. Haraway, *Staying with the Trouble*, 72.

96. Braverman, *Wild Life*, 44. See also Jamie Lorimer and Clemens Driessen, "Wild Experiments at the Oostvaardersplassen: Rethinking Environmentalism in the Anthropocene," *Transactions of the Institute of British Geographers* 39 (2014): 169–181.

97. Peter Sale (marine ecologist, professor emeritus, University of Windsor, Canada), written comments, August 7, 2017.

98. Colin Foord (cofounder and codirector, Coral Morphologic), e-mail communication, August 17, 2017.

99. This quotation and the others through the end of this chapter are from Foord, e-mail communication, August 17, 2017.

100. Anthony et al., "New Interventions Are Needed," 1421.

101. Foord, "A Hybrid Future."

INTERVIEW. RUTH GATES

1. Elizabeth Kolbert, "Unnatural Selection: What Will It Take to Save the World's Reefs and Forests?" *The New Yorker*, April 18, 2016. https://www.newyorker.com /magazine/2016/04/18/a-radical-attempt-to-save-the-reefs-and-forests.

2. Amanda Mascarelli, "Designer Reefs," *Nature* 508 (2014): 444–446.

CONCLUSION. CORAL SCIENTISTS ON THE BRINK

1. Anna L. Tsing, *The Mushroom at the End of the World: On the Possibility of Life in Capitalist Ruins* (Princeton University Press, 2015), 1.

2. Crochet Coral Reef, "About the Project," accessed July 5, 2017. http://crochet coralreef.org/about/index.php.

3. Donna Haraway, "Anthropocene, Capitalocene, Chthulucene: Staying with the Trouble," in *Anthropocene: Arts of Living on a Damaged Planet*, May 9, 2014. http:// opentranscripts.org/transcript/anthropocene-capitalocene-chthulucene/.

4. Haraway, "Anthropocene." See also Stefan Helmreich, "How Like a Reef: Figuring Coral, 1839–2010," accessed July 13, 2017. http://reefhelmreich.blogspot.com/. Helmreich explains that the Crochet Reef Project "can attune their human visitors and inquisitors to empirical and epistemological questions of scale and context—where context, drawing upon a once-upon-a-time literal but now more figural meaning, refers to a 'weaving together.'"

5. Donna J. Haraway, *Staying with the Trouble: Making Kin in the Chthulucene* (Duke University Press, 2016), 4.

6. Ibid., 64.

7. Coral Morphologic, "About," accessed July 5, 2017. http://www.coralmorphologic .com/about/.

8. Ibid.

9. Haraway, *Staying with the Trouble*, 72.

10. Stefan Helmreich, "How Scientists Think; about 'Natives', for Example. A Problem of Taxonomy among Biologists of Alien Species in Hawaii," *Journal of the Royal Anthropological Institute* 11 (2005): 107–128, at 124.

11. See figure 4.

12. Quoted in the Introduction.

13. See interchapter interview with Gates.

14. Haraway, *Staying with the Trouble*, 72.

15. Scott F. Gilbert, Jan Sapp, and Alfred I. Tauber, "A Symbiotic View of Life: We Have Never Been Individuals," *The Quarterly Review of Biology* 87 (2012): 325–341, at 333.

16. Haraway, *Staying with the Trouble*, 69.

17. Most coral scientists will agree that whatever nature used to be, it has irrevocably changed. Legal philosopher Jedediah Purdy articulates this mode of thinking:

"The natural and the artificial have merged at every scale. Climate change makes the global atmosphere, its chemistry and weather systems, into Frankenstein's monster— part natural, part made. The same is true of the seas, as carbon absorption turns the oceans acidic and threatens everything that lives in them." Jedediah Purdy, *After Nature: A Politics for the Anthropocene* (Harvard University Press, 2015), 15.

18. Late climate scientist Jerry Mahlman, quoted in Dale Jamieson, *Reason in a Dark Time: Why the Struggle against Climate Change Failed—and What It Means for Our Future* (Oxford University Press, 2014), 1.

19. Ken Nedimyer (founder and president, Coral Restoration Foundation), interview by author, telephone, October 17, 2017.

20. Jeremy Jackson, in Sandy Cannon-Brown, "From Despair to Repair 1080," vimeo video, posted July 1, 2014. https://vimeo.com/99653458 (transcribed by author).

21. Mike Hulme, "Reducing the Future to Climate: A Story of Climate Determinism and Reductionism," *Osiris* 26 (2011): 245–266.

22. Dipesh Chakrabarty, "The Politics of Climate Change Is More Than the Politics of Capitalism," *Theory, Culture & Society* 34(2–3) (2017): 25–37, at 27.

23. Ken Anthony et al., "New Interventions Are Needed to Save Coral Reefs," *Nature Ecology & Evolution* 1 (2017): 1420–1422, at 1420.

24. See, e.g., interchapter interview with ecologist Peter Sale.

25. Liz Burmester (Ph.D. student, Kaufman Lab, Boston University), interview by author, in person, Boston, MA, April 5, 2017.

26. Colin Foord (cofounder, Coral Morphologic), e-mail communication, August 17, 2017. Matthew Chrulew similarly highlights the tragic irony that "making live" through captive breeding and assisted reproduction results in a death sentence for the particular nonhumans involved. Matthew Chrulew, "Managing Love and Death at the Zoo: The Biopolitics of Endangered Species Preservation," *Australian Humanities Review* 50 (2011): 137–157.

27. Pamela M. Lee, *Chronophobia: On Time in the Art of the 1960s* (MIT Press, 2004).

28. Quoted in ibid., 264.

29. Ibid., xxv.

30. Science and Security Board, Bulletin of the Atomic Scientists, *It Is Two and a Half Minutes to Midnight: 2017 Doomsday Clock Statement*, edited by John Mecklin, 2017. https://thebulletin.org/sites/default/files/Final%202017%20Clock%20Statement.pdf.

31. Ruth Gates (director, Hawai'i Institute of Marine Biology; president, International Society for Reef Studies), interview by author, in person, July 1, 2016.

32. Baruch Rinkevich (senior scientist, National Institute of Oceanography, Haifa, Israel), interview by author, telephone, May 1, 2017.

33. *Chasing Coral*, a film produced by Larissa Rhodes and directed by Jeff Orlowski (2017, Exposure Labs, Netflix).

34. See chapter 2.

INTERVIEWS

Anuenue Fisheries Research Center Coral Nursery, Honolulu, HI. Observation, June 30, 2016.

Arnaud-Haond, Sophie. French Research Institute for Exploitation of the Sea (IFRE-MER–UMR), Marine Biodiversity, Exploitation and Conservation research unit. Skype, May 13, 2016.

Barandiaran, Mike. Manager, Vieques Refuge, U.S. Fish and Wildlife Service. In person, Vieques, PR, January 12–13 and 16, 2015.

Baums, Iliana B. Associate professor of biology, The Pennsylvania State University, and director, Center for Marine Science and Technology. Skype, September 17, 2015.

Biggs, Duan. Postdoctoral research fellow, Australian Research Council Centre of Excellence for Environmental Decisions, Queensland, Australia. Skype, May 7, 2015.

Bramanti, Lorenzo. Marine biologist, French National Centre for Scientific Research (CNRS-LECOB). In person, Buffalo, NY, March 16, 2017; Skype, March 21, 2017.

Burmester, Liz. Ph.D. student, Kaufman Lab. In person, Boston University, Boston, MA, April 5, 2017.

Byrne, Maria. Professor of marine and developmental biology, University of Sydney. In person, University of Sydney, Sydney, Australia, May 10, 2015.

Camhi, Merry. Director, New York Seascape Policy Program, Wildlife Conservation Society. Telephone, joint interview with Noah Chesnin, December 7, 2015.

Chesnin, Noah. Manager, New York Seascape Policy Program, Wildlife Conservation Society. Telephone, joint interview with Merry Camhi, December 7, 2015.

Chika-Suey, Katia, and Jenna Budke. Ph.D. students in marine biology. In person and observations at Waiʻopae tide pools, HI, June 27, 2016.

Cinner, Joshua. Professorial research fellow and chief investigator, Australian Research Council Centre of Excellence for Coral Reef Studies. In person, Waikiki, HI. June 21, 2016.

Coconut Island Coral Nursery (with director Ruth Gates and staff). Observation, Oahu, HI, July 1, 2016.

Coffroth, Mary Alice. Professor of geology, Graduate Program in Evolution, Ecology and Behavior, University at Buffalo, SUNY. In person and observations, Buffalo, NY, August 31 and November 16, 2015; joint interview with Howard Lasker, August 19, 2015.

Day, Jon. Former director of conservation, Great Barrier Reef Marine Park Authority; Australian Research Council Centre of Excellence for Coral Reef Studies. Skype, May 22, 2015.

Dunbabin, Matthew. Principal research fellow, Department of Science and Engineering, Queensland University of Technology, Australia. Skype, January 20, 2016.

Eakin, Mark. Coordinator, NOAA Coral Reef Watch. Telephone, November 16, 2015; e-mail communication, October 20, 2017.

Finnerty, John. Associate professor of biology, Boston University. In person, Boston University, Boston, MA, April 5, 2017.

Fogarty, Nicole. Assistant professor, Nova Southeastern University Oceanographic Center. In person, Buffalo, NY, April 26, 2017; telephone, May 1, 2017; Skype, June 15, 2017; e-mail communications, April–August 2017.

Foord, Colin. Cofounder and codirector, Coral Morphologic; telephone and e-mail communications, July 2017–January 2018.

Forsman, Zac H. Researcher, Hawai'i Institute of Marine Biology, and coral specialist, State of Hawai'i. Telephone, March 1, 2016; in person, Waikiki and Honolulu, HI, June 20 and 30, 2016; e-mail communications, June–August 2016.

Frias-Torres, Sarah. Marine ecologist, Smithsonian Marine Station, Fort Pierce, Florida. In person, International Coral Reef Symposium, Oahu, HI, June 23, 2016.

Gates, Ruth. Director, Hawai'i Institute of Marine Biology, and president, International Society for Reef Studies. Skype, January 25, 2016; in person, Coconut Island, Oahu, HI, July 1, 2016; e-mail communications, 2016–2017.

Genin, Amatzia. Professor of ecology, evolution and behavior, and scientific director, Interuniversity Institute for Marine Sciences in Eilat; in person, Eilat, Israel, June 11, 2015.

Gjerde, Kristina. Adjunct professor, Middlebury Institute of International Studies at Monterey, and senior high seas advisor, International Union for Conservation of Nature (IUCN). Skype, June 7, 2016; observations, UNCLOS PrepCom, United Nations Headquarters, New York, NY, August 30, 2016.

Graham, Alistair. High seas policy advisor, World Wildlife Fund International, and NGO nominee on the Australian government delegation, UNCLOS PrepCom. In person, United Nations Headquarters, New York, NY, August 30, 2016.

Hagedorn, Mary. Director, MarineGEO Hawai'i, Smithsonian Conservation Biology Institute/Hawai'i Institute of Marine Biology. Telephone, September 24, 2015; August 2016.

Halpern, Ben. Professor of marine ecology, University of California, Santa Barbara. In person, International Coral Reef Symposium, Oahu, HI, June 23, 2016.

Henry, Lea-Anne. Coral reef scientist, research fellow in the School of Life Sciences, Heriot Watt University. In person, Edinburgh, United Kingdom, April 20, 2016.

Hernández-Delgado, Edwin A. Marine biologist and affiliate researcher and lecturer, Department of Biology, University of Puerto Rico, Río Piedras. In person, Culebra, PR, January 17, 2015; telephone, January 27, 2015.

Hoegh-Guldberg, Ove. Director, Global Change Institute; professor of marine science, University of Queensland, Brisbane, Australia. Skype, February 25, 2015, and May 22, 2017; in person, Waikiki, HI, June 23, 2016.

Jackson, Jeremy. Professor, Scripps Institution of Oceanography, and senior scientist emeritus, Smithsonian Tropical Research Institute in the Republic of Panama. Skype, February 22 and March 16, 2017; e-mail communications, October 2017.

Janse, Max. Ocean curator, Royal Burgers' Zoo, The Netherlands. Skype, October 30, 2015.

Jaspars, Marcel. Professor of chemistry, University of Aberdeen, and chief scientific officer, Ripptide Pharma, Aberdeen, United Kingdom. In person, New York, NY, August 29, 2016.

Jiménez-Marrero, Nilda. Biologist, Puerto Rico Department of Natural Resources. In person, Boquerón, PR, January 20, 2014; Telephone, October 7, 2014.

Johnson, David. Director, Seascape Consultants Ltd.; coordinator, Global Ocean Biodiversity Initiative; and former executive secretary, OSPAR Commission. Skype, joint interview with Philip Weaver, May 26, 2016.

Jones, Daniel. Researcher, National Oceanography Centre. Skype, April 10, 2017.

Kaufman, Les. Professor of biology, Boston University. In person, Waikiki, Oahu, HI, June 22, 2016; Boston University, Boston, MA, April 5, 2017.

Kelsey, Elin. Associate faculty, School of Environment and Sustainability, Royal Roads University. Skype, July 22, 2016; in person, Munich, Germany, July 3, 2017.

Khatib, Oussama. Director, Stanford Robotics Lab. Skype, May 31, 2016.

Kleypas, Joanie. Marine ecologist and geologist, National Center for Atmospheric Research, Climate and Global Dynamics. In person, International Coral Reef Symposium, Oahu, HI, June 23, 2016.

Klobuchar, Rick. Associate curator, The Florida Aquarium. In person, Tampa, FL, December 28, 2015.

Knowlton, Nancy. Sant Chair in Marine Science, Smithsonian's National Museum of Natural History. Skype, July 11, 2016.

Lankari, Eli. Deputy mayor, Eilat. In person, Eilat, Israel, January 8, 2017.

Lasker, Howard. Professor of geology and director, Graduate Program in Evolution, Ecology and Behavior, University at Buffalo, SUNY. In person, Buffalo, NY, on three dates: August 17, 2014; April 13, 2015; and December 15, 2016; joint interview with Mary Alice Coffroth, August 19, 2015.

Lewis, Cindy. Research associate, Keys Marine Lab, Florida. Telephone, September 8, 2015.

Lilyestrom, Craig. Director, Marine Resources Division, Puerto Rico Department of Natural and Environmental Resources (DNER). Telephone, November 12, 2014.

Lucking, Mary Ann. Director, CORALations. In person, Culebra, PR, January 17, 2015.

Machlis, Gary. Professor of environmental sustainability, Clemson University, and science advisor to the director, National Park Service. Telephone, September 28, 2014.

Matz, Mikhail. Professor of integrative biology, University of Texas at Austin. Skype, May 2, 2016.

McKnight, Margo. Vice president of biological operations, The Florida Aquarium. In person, Tampa, FL, December 28, 2015.

Miller, Margaret W. Research ecologist, NOAA/National Marine Fisheries Service, Southeast Fisheries Science Center. Telephone, September 16, 2015; in person, Honolulu, HI, June 21, 2016; e-mail communication, July 18, 2016.

Moore, Jennifer. Marine conservation program manager, NOAA. Telephone, September 30, 2015.

Moore, Tom. Coral restoration program manager, NOAA Restoration Center. Telephone, October 2 and 7, 2015.

Mott, Bill. Director, The Ocean Project. Telephone, January 12, 2016.

Moulding, Alison. Natural resource specialist, Fisheries Southeast Regional Office Protected Resources Division, NOAA. Telephone, December 16, 2015.

Navarro-Delgado, Cruz. Department of Biostatistics and Epidemiology, School of Public Health, University of San Juan, PR. Telephone, September 18, 2014.

Nedimyer, Ken. Founder and president, Coral Restoration Foundation. Telephone, January 4, 2016, and October 17, 2017; in person, June 24, 2016; e-mail communication, October 13, 2017.

NOAA Workshop, Hilo, HI. Observation, June 28, 2016.

Olson, Julia. Executive director and chief legal counsel, Our Children's Trust. Telephone, November 6, 2017.

Otano, Abimarie. Graduate student, University of Puerto Rico. In person, Culebra, PR, January 17, 2015.

Page, Christopher A. Staff biologist, Coral Reef Restoration, Mote Marine Laboratory and Aquarium. Telephone, September 11, 2015.

Palumbi, Stephen (Steve). Professor of biological sciences and director of Hopkins Marine Station, Stanford University. In person, Waikiki, HI, June 24, 2016.

Pichon, Michel. Tropical marine consultant and honorary research associate, Museum of Tropical Queensland, Australia. In person, Waikiki, HI, June 20, 2016.

Penfold, Linda. Director, South-East Zoo Alliance for Reproduction and Conservation (SEZARC). Skype, July 26, 2016.

Petersen, Dirk. Director, SExual COral REproduction (SECORE). Skype, September 23, 2015, and January 6, 2016; in person, Waikiki, HI, June 20, 2016.

Porter, James W. Josiah Meigs Distinguished Professor, Odum School of Ecology, University of Georgia. Telephone, September 24, 2014.

Possingham, Hugh. Professor of mathematics and ecology, The University of Queensland, Australia; director, Australian Research Council Centre of Excellence for Environmental Decisions; and director, National Environmental Research Program Environmental Decisions Hub. Skype, May 5, 2015.

Quintana, Ángela Martinez. Ph.D. student, University at Buffalo, SUNY. In person, October 2016; e-mail communication, June 17, 2017.

Rahn, Jennifer. Associate professor of geography, Samford University. In person, American Association of Geographers Annual Meeting, Boston, MA, April 8, 2017.

Rhyne, Andrew (Andy). Associate professor of marine biology, Roger Williams University. In person, New England Aquarium, Boston, MA, May 11, 2016, joint interview with Michael Tlusty.

Rinkevich, Baruch (Buki). Senior scientist, National Institute of Oceanography, Haifa, Israel. Telephone, June 18, 2015; March 23, 2017; and May 1, 2017; in person, Waikiki, HI, June 21–22, 2016; e-mail communications, 2016–2018.

Roberts, J Murray. Professor of marine biology, The University of Edinburgh. Skype, January 28, 2016; in person, Edinburgh, United Kingdom, April 20, 2016; e-mail communication, July 3, 2017.

Rodrigues, Daniel. Remedial park manager (Vieques), U.S. Environmental Protection Agency. In person, Isabelle Segundo, Vieques, PR, January 16, 2015.

Sale, Peter. Marine ecologist, professor emeritus, University of Windsor, Canada. Skype, July 7, 2016; e-mail communications, July–September 2017; written comments, August 7, 2017.

Satashita, Miyoko. Oceans director and senior counsel, Center for Biological Diversity. Skype, October 8, 2015.

Schick, Mark. Special exhibits collection manager, Fishes Department, John G. Shedd Aquarium. Telephone, April 26, 2016.

Shaham, Assaf. Founder and CEO, OkCoral coral farm, Ein Yahav, Israel. In person and observations, Ein Yahav, Israel. June 8, 2015.

Shashar, Nadav. Professor of life sciences, Ben-Gurion University of the Negev. In person, Eilat, Israel, January 8, 2017.

Shefi, Dor. Ph.D. student, Ben-Gurion University of the Negev. In person, Eilat, Israel, June 8, 2017.

Sheppard, Bart. Director, Steinhart Aquarium, San Francisco, California. Telephone, October 1, 2015.

Silander, Susan. Project leader, Caribbean National Wildlife Refuge Complex, U.S. Fish and Wildlife Service. Telephone, August 24, 2014; in person, Vieques, PR, January 12–13, 2015.

Silverman, Jack. Scientist, National Institute of Oceanography, Haifa, Israel. Skype, June 1, 2015.

Skirving, William. NOAA Coral Reef Watch, Kirwan, Queensland, Australia. In person, NOAA workshop, University of Hawai'i at Hilo, HI, June 28, 2016.

Smith, Walt. Founder and director, Walt Smith International Aquaculture, Fiji. Skype, July 18, 2016.

Soriano, Maricor. Associate professor of physics, National Institute of Physics, University of the Philippines, Diliman. In person, Oahu, HI, June 22, 2016.

Taggar, Shmulik. Former head of tourism department, Eilat Municipality. In person, Eilat, Israel, June 8, 2017.

Takabayashi, Misaki. Associate dean, College of Arts and Sciences, and professor of marine science, University of Hawai'i at Hilo. In person, Hilo, HI, June 28, 2016.

Than, John. Manager, Center for Conservation and Collection, Florida Aquarium. In person, Tampa, FL, December 28, 2015.

Thornhill, Dan. Program director, Biological Oceanography, Auburn University. In person, Buffalo, NY, April 13, 2016.

Tlusty, Michael. Director of research, New England Aquarium. In person, New England Aquarium, Boston, MA, joint interview with Andrew Rhyne, May 11, 2016.

Tortorici, Cathy. Division chief, Endangered Species Act Interagency Cooperation Division, NOAA. Telephone, December 16, 2015.

Tudhope, Sandy. Professor, School of GeoSciences, University of Edinburgh, United Kingdom. Skype, January 29, 2016.

van Oppen, Madeleine J. H. Professor, University of Melbourne, and senior principal research scientist, Australian Institute of Marine Science. Skype, October 15, 2015.

Vera, Juan. State underwater archaeologist, Puerto Rico. In person, Isabelle Segundo, Vieques, PR, January 16, 2015.

Veron, John "Charlie." Former chief scientist, Australian Institute of Marine Science. In person, Waikiki, HI, June 22, 2016.

Vevers, Richard. Founder and CEO, The Ocean Agency. Skype, August 5 and 6, 2016.

Voolstra, Christian (Chris) R. Associate professor of marine science, Biological and Environmental Science and Engineering Division, King Abdullah University of Science and Technology. Skype, May 12, 2016; in person, Oahu, HI, June 20–22, 2016; e-mail communications, summer 2016.

Wachenfeld, David. Director of reef recovery, Great Barrier Reef Marine Park Authority. Skype, January 4, 2016.

Weaver, Phillip. Managing director, Seascape consultants LTD; scientific coordinator, Global Ocean Biodiversity Initiative; and former deputy director, National Oceanography Centre, Southampton, United Kingdom. Skype, joint interview with David Johnson, May 26, 2016.

Wolf, Shaye. Climate change director, Center for Biological Diversity. Skype, November 4, 2015.

Zvuloni, Assaf. Director, Eilat Marine Reserve. In person and observations, Eilat, Israel, June 12 and 13, 2015.

INDEX